国土资源部行业基金项目(201311062)的研究成果之一

宜昌地区赫南特动物群及其生境和灭绝原因以及兰多维列世生物群演变

Hirnantia Fauna, its habitat and extinctive cause of Middle Hirnantian and biota evolution of Llandoverian in Yichang District, Western Hubei, Central China

曾庆銮　陈孝红　王传尚
张　淼　韩会卿　彭中勤　著

内容简介

该书一是对宜昌地区五峰组观音桥段赫南特动物群化石的详细采集和描述，共有该动物群15个科(含2个新科)、35个属(含6个新属)、43个种(含11个新种)，附有60个精美化石图版、29张插图，并对其中5个老属的特征进行补充和修订。对该动物群灭绝的原因提出了新的观点，并对宜昌地区该动物群的生态环境提出了新的见解。首次在宜昌地区观音桥段内发现浊流沉积层序，并指出宜昌地区观音桥段存在不完整性。首次在宜昌地区观音桥段与龙马溪组黑色页岩段之间的转换分界面上发现该动物群成员与 *persculptus* 带笔石群成员的化石共同保存的情况。二是对宜昌地区兰多维列世前后发生过两幕海侵与海退、营造了巨大差别的古生态环境和导致先后生物群发生强烈演变进行讨论，对其中腕足类的一些新材料进行描述，共有3个目、3个科、4个属(含2个新属)、2个亚属(含1个新亚属)、6个种(含3个新种)。

本书可供地质、地层、古生物、古生态、古地理、地史、古气候、石油与天然气等行业的生产、教学和科研工作者参考；也可作为对大、中学生进行地球发展史科普教育的材料。

图书在版编目(CIP)数据

宜昌地区赫南特动物群及其生境和灭绝原因以及兰多维列世生物群演变/曾庆銮,陈孝红,王传尚,张淼,韩会卿,彭中勤著．—武汉：中国地质大学出版社,2016.6
ISBN 978-7-5625-3835-6

Ⅰ.①宜…
Ⅱ.①曾…②陈…③王…④张…⑤韩…⑥彭…
Ⅲ.①生物群-动物区系-宜昌市
Ⅳ.①Q915.763.3

中国版本图书馆 CIP 数据核字(2016)第152686号

宜昌地区赫南特动物群及其生境和灭绝原因以及兰多维列世生物群演变	曾庆銮　陈孝红　王传尚 张　淼　韩会卿　彭中勤	著

责任编辑：胡珞兰　龙昭月	组　稿：张晓红	责任校对：戴莹
出版发行：中国地质大学出版社(武汉市洪山区鲁磨路388号)		邮政编码：430074
电　话：(027)67883511	传　真：67883580	E-mail:cbb@cug.edu.cn
经　销：全国新华书店		http://www.cugp.cug.edu.cn
开本：880毫米×1230毫米 1/16	字数：390千字　印张：7.75　图版：70	
版次：2016年6月第1版	印次：2016年6月第1次印刷	
印刷：湖北睿智印务有限公司	印数：1—1 000册	
ISBN 978-7-5625-3835-6		定价：68.00元

如有印装质量问题请与印刷厂联系调换

前　言

1　赫南特动物群的研究概况

赫南特动物群(Hirnantia Fauna)一名由 Temple(1965)首先提出。在我国该腕足动物群广泛分布于上扬子区,并且只产在上奥陶统顶部赫南特阶(Hirnantian)中部,即五峰组观音桥段(层)内。我国对该腕足动物群进行研究相对较晚,起初也仅有零星报道(戎嘉余等,1974;曾庆銮,1977;阎国顺,1978)。直到 20 世纪 70 年代末,戎嘉余(1979)才对中国赫南特动物群进行全面、深入的研究,为我国对赫南特动物群的研究打下了坚实的基础。嗣后常美丽(1981)、曾庆銮(1983)、Rong Jiayu(1984)、曾庆銮(1987)、戎嘉余(2006)等先后又对宜昌地区赫南特动物群进行过较详细的研究。

这次的研究就是在上述研究的基础上进行的,并已经取得了重大的进展。其中有以下 6 个方面是值得一提的:一是对赫南特动物的研究精度向前推进了一大步,使赫南特动物群的组合面貌焕然一新,其中在扭月贝超科内取得了更令人注目的新进展。目前在宜昌地区获得赫南特动物群共有 15 个科(含 2 个新科)、35 个属(含 6 个新属)、43 个种(含 11 个新种),共贴 60 个图版(即图版 1~60),而且特征都很清晰。同时指出宜昌地区的赫南特动物群应是接近于暖水型的,而不是冷水型的。另外,还对 5 个老属的特征进行了修订。目前获得如此庞大的赫南特动物群,这在当今世界是从未有过的。二是对赫南特动物群的生态环境提出了新的见解。三是对赫南特动物群灭绝的原因提出了新的观点。四是在宜昌地区观音桥段内首次发现浊流沉积层序,并首次提出观音桥段(层)存在着不完整性。五是在宜昌地区观音桥段与龙马溪组黑色页岩段的转换分界面上首次发现赫南特动物群成员的化石与龙马溪组黑色页岩段最底部 *Normalograptus persculptus* 带笔石群成员的化石共同保存在转换面的情况。六是简要地论述在地史长河中距今 443.7Ma 之前,并且只有经历约 200 万年(2Myr±)的赫南特阶(Hirnantian)在宜昌地区先后发生一连串的地质事件与著名的冈瓦纳大陆冰盖的形成→冰盖向大陆架扩张→冰盖从大陆架撤退→大陆冰盖消融等演变阶段之间的紧密关系。

2　赫南特动物群的化石材料

本书赫南特动物群(Hirnantia Fauna)的化石材料是采集于黄陵背斜东翼、宜昌市以北沿着去保康县公路 22~42km 之间的黄花场(厚 30cm,化石代号 HH)、丁家坡(厚 18cm,化石代号 DH)、王家湾(厚 20cm,化石代号 WH)3 条剖面的五峰组观音桥段(或称观音桥层)。我们除在野外实地直接采集之外,还从野外采回约 0.3t 观音桥段的大块岩石,在室内经过近 8 个月的时间将大块岩石劈开,并在双目镜下修理,从 5000 多枚标本中精选出 682 枚完美的化石标本拍摄成化石照片。宜昌地区赫南特动物群的化石极为丰富,不仅保存很好,而且分异度极大,目前共获得 8 个目、12 个超科、15 个科(含 2 个新科)、35 个属(含 6 个新属)、43 个种(含 11 个新种),共贴 60 个图版。获得如此庞大的赫南特动物群,这在世界各地是未曾有过的。其中以扭月贝目的分异度最大,目前共有 13 个属,占整个赫南特动物群 35 个属的 37.1%;次之为正形贝目,共有 12 个属,占 34.1%;而其他各个目的数量都较少,它们分别为直形贝目 4 个属,占 11.4%;舌形贝亚门 4 个属,占 11.4%;小嘴贝目 1 个属,占 3%;无窗贝目 1 个属,占 3%。上述是宜昌地区赫南特动物群成员组合的概况;若以化石丰度来看,则以正形贝目当中的德姆贝亚目各成员的化石最为丰富,而扭月贝目各成员的化石虽然也很丰富,但远不如德姆贝亚目各成员的丰度。

3 宜昌地区兰多维列世生物群演变

宜昌地区在兰多维列世早期发生第一幕大海侵,其所引来的为冷水型的海水,当时的生态底域位置处在 BA6,只有在表层水生存着极为昌盛的浮游笔石动物群。其后,宜昌地区发生兰多维列世第一幕大海退事件,推断此事件应与黔北地区"桐梓上升"事件为同一地史时期的地质事件。之后,宜昌地区在晚兰多维列世早期又发生第二幕大海侵,其所引来的海水为暖水型的,其古生态环境与其前期发生了质的变化,营造了极为适宜于底栖动物生存与繁衍的一种生态环境,导致了沉积罗惹坪组时底栖生物群大复苏。本书仅对其中腕足类的一些新材料进行描述,它们共有 3 个目、3 个科、4 个属(含 2 个新属)、2 个亚属(含 1 个新亚属)、6 个种(含 3 个新种),并贴 5 个图版(即图版 64 至图版 68)和 4 张插图(即插图 35 至插图 38)。再其后,宜昌地区又发生了兰多维列世第二幕大海退,即产生了"扬子上升"地质事件,最后导致上扬子海走向消亡。

4 致谢

国土资源部行业基金项目(201311062)为这一研究工作提供了资助;武汉地质调查中心古生物室为这一研究工作提供了方便与支持;汪啸风、孟繁松两位教授参加野外工作;杨振强教授对文中浊流沉积进行了指点;武汉地质调查中心宜昌长江地质灾害防治工程勘察设计院(地质灾害研究室)为这一研究工作提供了方便与帮助,刘辉帮助清绘图表、文稿打字。笔者在此一并致以诚挚的感谢!

由于作者水平有限,书中难免存在不足和错误,敬请读者批评指正。

著 者
2015 年 12 月

目　录

第1章　地层和时代以及化石保存形式 ……………………………………………………… (1)
1.1　地层 ……………………………………………………………………………………… (1)
1.2　时代 ……………………………………………………………………………………… (3)
1.3　化石保存形式 …………………………………………………………………………… (3)

第2章　浊流沉积及观音桥段的不完整性 …………………………………………………… (4)
2.1　浊流沉积 ………………………………………………………………………………… (4)
2.1.1　王家湾观音桥段浊流沉积剖面 …………………………………………………… (4)
2.1.2　丁家坡观音桥段浊流沉积剖面 …………………………………………………… (5)
2.1.3　浊流依据与成因 …………………………………………………………………… (5)
2.2　观音桥段的不完整性 …………………………………………………………………… (6)

第3章　赫南特动物群的生境和地质事件 …………………………………………………… (7)
3.1　地质事件 ………………………………………………………………………………… (7)
3.2　晚凯迪早期的生态底域位置及其生态类型 …………………………………………… (7)
3.2.1　发生时间 …………………………………………………………………………… (7)
3.2.2　标志物和生态底域位置 …………………………………………………………… (7)
3.3　晚凯迪晚期至早赫南特期的生态底域位置及其生态类型 ………………………… (9)
3.3.1　发生时间 …………………………………………………………………………… (9)
3.3.2　标志物和生态底域位置 …………………………………………………………… (9)
3.3.3　古生态环境 ………………………………………………………………………… (9)
3.4　赫南特动物群的生态环境 ……………………………………………………………… (9)
3.4.1　赫南特动物群生存的时间 ………………………………………………………… (9)
3.4.2　标志物及生态底域位置 …………………………………………………………… (9)
3.4.3　两大地质事件及赫南特动物群的生态环境 …………………………………… (10)

第4章　赫南特动物群灭绝原因的探讨 ……………………………………………………… (11)
4.1　灭绝时间 ………………………………………………………………………………… (11)
4.2　标志物和生态底域位置 ………………………………………………………………… (11)
4.2.1　标志物 ……………………………………………………………………………… (11)
4.2.2　生态底域位置变迁 ………………………………………………………………… (11)
4.3　赫南特动物群灭绝原因 ………………………………………………………………… (11)

第5章　宜昌地区兰多维列世生物群演变 …………………………………………………… (13)
5.1　沉积龙马溪组时的动物群 ……………………………………………………………… (13)
5.1.1　早期古生态环境和动物群 ………………………………………………………… (13)
5.1.2　晚期古生态环境和动物群 ………………………………………………………… (13)
5.2　沉积罗惹坪组时生物群大复苏 ………………………………………………………… (15)
5.2.1　沉积罗惹坪组时的古生态环境 …………………………………………………… (15)

5.2.2 沉积罗惹坪组时生物群大复苏 ……………………………………………………………… (15)
5.3 上扬子海走向消亡 ………………………………………………………………………………… (16)
5.3.1 沉积物和接触关系 ………………………………………………………………………… (16)
5.3.2 上扬子海走向消亡的证据 ………………………………………………………………… (17)

第6章 化石系统描述 ……………………………………………………………………………… (18)

6.1 赫南特动物群的化石 …………………………………………………………………………… (18)

腕足动物门 Brachiopoda Duméril,1806 ……………………………………………………………… (18)
舌形贝亚门 Linguliformea Williams et others,1996 ………………………………………………… (18)
 舌形贝纲 Lingulata Gorjansky et Popov,1985 ……………………………………………………… (18)
  舌形贝目 Lingulida Waagen,1885 ………………………………………………………………… (18)
   平圆贝超科 Discinoidea Gray,1840 ……………………………………………………………… (18)
    平圆贝科 Discinidae Gray,1840 ………………………………………………………………… (18)
     圆凸贝属 *Orbiculoidea* Orbigny,1847 ………………………………………………………… (18)
骨髅贝亚门 Craniiformea Popov et others,1993 …………………………………………………… (19)
 骨髅贝纲 Craniata Williams et others,1996 ………………………………………………………… (19)
  非骨髅贝目 Craniopsida Gorjansky et Popov,1985 ………………………………………………… (19)
   非骨髅贝超科 Craniopsoidea Williams,1963 ……………………………………………………… (19)
    非骨髅贝科 Craniopsidae Williams,1963 ………………………………………………………… (19)
     非骨髅贝属 *Craniops* Hall,1859b ……………………………………………………………… (19)
骨髅贝目 Craniida Waagen,1885 ……………………………………………………………………… (20)
 骨髅贝超科 Cranioidea Menke,1828 ………………………………………………………………… (20)
  骨髅贝科 Craniidae Menke,1828 …………………………………………………………………… (20)
   刺骨髅贝属 *Acanthocrania* Williams,1943 …………………………………………………… (20)
   友基贝属 *Philhedra* Koken,1889 ……………………………………………………………… (21)
小嘴贝亚门 Rhynchonelliformea Williams et others,1996 ………………………………………… (21)
 扭月贝纲 Strophomenata Williams et others,1996 ………………………………………………… (21)
  扭月贝目 Strophomenida Öpik,1934 ………………………………………………………………… (21)
   扭月贝超科 Strophomenoidea King,1846 ………………………………………………………… (21)
    扭月贝科 Strophomenidae King,1846 …………………………………………………………… (21)
     叉形贝亚科 Furcitellinae Williams,1965 ……………………………………………………… (21)
      小月贝属(新属) *Minutomena* Zeng,Zhang et Han(gen. nov.) ………………………… (21)
    瑞芬贝科 Rafinesquinidae Schuchert,1893 ……………………………………………………… (23)
     薄皱贝亚科 Leptaeninae Hall et Clarke,1894 ………………………………………………… (23)
      薄皱贝属 *Leptaena* Dalman,1828 …………………………………………………………… (23)
      薄盖贝属 *Leptaenopoma* Marek et Havlíček,1967 ………………………………………… (25)
    雕月贝科 Glyptomenidae Williams,1965 ………………………………………………………… (29)
     雕月贝亚科 Glyptomeninae Williams,1965 …………………………………………………… (29)
      平月贝属 *Paromalomena* Rong,1979 ……………………………………………………… (29)
    中华月贝科(新科) Sinomenidae Zeng,Chen et Zhang(fam. nov.) ……………………………… (30)
     隐月贝属 *Aphanomena* Bergström,1968 …………………………………………………… (31)
     始齿扭贝属 *Eostropheodonta* Bancroft,1949 ……………………………………………… (32)
     宜昌月贝属(新属) *Yichangomena* Zeng,Zhang et Han(gen. nov.) ……………………… (34)

中华月贝属（新属）*Sinomena* Zeng,Chen et Zhang(gen. nov.) …………………………… (36)
湖北月贝属（新属）*Hubeinomena* Zeng,Chen et Zhang(gen. nov.) ………………………… (37)
褶脊贝超科 Plectambonitoidea Jones,1928 ………………………………………………………… (38)
  异脊贝科 Xenambonitidae Cooper,1956 ………………………………………………………… (38)
    埃月贝亚科 Aegiromeninae Havliček,1961 …………………………………………………… (38)
      埃月贝属 *Aegiromena* Havliček,1961 ……………………………………………………… (38)
      埃吉贝属 *Aegiria* Öpik,1933 ………………………………………………………………… (41)
      似戟贝属 *Chonetoidea* Jones,1928 ………………………………………………………… (43)
      三板月贝属（新属）*Trimena* Zeng, Wang et Peng(gen. nov.) …………………………… (45)
直形贝目 Orthotetida Waagen,1884 ………………………………………………………………… (46)
  直形贝亚目 Orthotetidina Waagen,1884 ………………………………………………………… (46)
    直形贝超科 Orthotetoidea Waagen,1884 ……………………………………………………… (46)
      法顿贝科 Fardeniidae Williams,1965 ………………………………………………………… (46)
        法顿贝亚科 Fardeniinae,Williams,1965 …………………………………………………… (46)
          法顿贝属 *Fardenia* Lamont,1935 ………………………………………………………… (46)
          库林贝属 *Coolinia* Bancroft,1949 ………………………………………………………… (47)
  三重贝亚目 Triplesiidina Moore,1952 …………………………………………………………… (48)
    三重贝超科 Triplesioidea Schuchert,1913 …………………………………………………… (48)
      三重贝科 Triplesiidae Schuchert,1913 ……………………………………………………… (48)
        三重贝属 *Triplesia* Hall,1859 ……………………………………………………………… (48)
        克利夫通贝属 *Cliftonia* Foerste,1909 …………………………………………………… (50)
小嘴贝纲 Rhynchonellata Williams et others,1996 ……………………………………………… (53)
  正形贝目 Orthida Schuchert et Cooper,1932 …………………………………………………… (53)
    正形贝亚目 Orthidina Schuchert et Cooper,1932 …………………………………………… (53)
      褶正形贝超科 Plectorthoidea Schuchert et LeVene,1929 ………………………………… (53)
        弓正形贝科 Toxorthidae Rong,1984 ………………………………………………………… (53)
          弓正形贝属 *Toxorthis* Temple,1968 ……………………………………………………… (53)
    德姆贝亚目 Dalmanellidina Moore,1952 ……………………………………………………… (54)
      德姆贝超科 Dalmanelloidea Schuchert,1913 ……………………………………………… (54)
        德姆贝科 Dalmanellidae Schuchert,1913 ………………………………………………… (54)
          德姆贝亚科 Dalmanellinae Schuchert,1913 ……………………………………………… (54)
            德姆贝属 *Dalmanella* Hall et Clarke,1892 …………………………………………… (54)
            安尼贝属 *Onniella* Bancroft,1928 ……………………………………………………… (56)
            特鲁西贝属 *Trucizetina* Havliček,1974 ……………………………………………… (58)
            奇异正形贝属 *Mirorthis* Zeng,1983 …………………………………………………… (61)
            似奇异正形贝属（新属）*Paramirorthis* Zeng, Wang et Peng(gen. nov.) ………… (62)
        全形贝超科 Enteletoidea Waagen,1884 …………………………………………………… (64)
          德拉勃贝科 Draboviidae Havliček,1950 ………………………………………………… (64)
            德拉勃贝亚科 Draboviinae Havliček,1950 ……………………………………………… (64)
              德拉勃贝属 *Drabovia* Havliček,1950 ………………………………………………… (64)
              小德拉勃贝属 *Drabovinella* Havliček,1950 ………………………………………… (66)
              赫南特贝属 *Hirnantia* Lamont,1935 ………………………………………………… (68)
              辛奈贝属 *Kinnella* Bergström,1968 ………………………………………………… (74)

德拉勃正形贝属 *Draborthis* Marek et Havlíček,1967	(77)
难得正形贝科(新科) Dysprosorthidae Zeng et Zhang(fam. nov.)	(80)
难得正形贝属 *Dysprosorthis* Rong,1984	(80)
小嘴贝目 Rhynchonellida Kuhn,1949	(82)
孔嘴贝超科 Rhynchotrematoidea Schuchert,1913	(82)
三角嘴贝科 Trigonirhynchiidae Schmidt,1965	(82)
嘴室贝亚科 Rostricellulinae Rozman,1969	(82)
小褶窗贝属 *Plectothyrella* Temple,1965	(82)
无窗贝目 Athyridida Boucot,Johnson et Staton,1964	(84)
无窗贝亚目 Athyridina Boucot,Johnson et Staton,1964	(84)
小双分贝超科 Meristelloidea Waagen,1883	(84)
小双分贝科 Meristellidae Waagen,1883	(84)
小双分贝亚科 Meristellinae Waagen,1883	(84)
欣德贝属 *Hindella* Davidson,1882	(84)

6.2 罗惹坪期一些腕足类的新材料 ...... (88)

扭月贝目 Strophomenida Öpik,1934	(88)
褶脊贝超科 Plectambonitoidea Jones,1928	(88)
准小薄贝科 Leptellinidae Ulrich et Cooper,1936	(88)
准小薄贝亚科 Leptellininae Ulrich et Cooper,1936	(88)
准小薄贝属 *Leptellina* Ulrich et Cooper,1936	(88)
小墨西贝亚属 *Leptellina*(*Merciella*) Lamont et Gilbert,1945	(88)
似小薄贝属 *Leptelloidea* Jones,1928	(89)
正形贝目 Orthida Schuchert et Cooper,1932	(90)
正形贝亚目 Orthidina Schuchert et Cooper,1932	(90)
褶正形贝超科 Plectorthoidea Schuchert et LeVene,1929	(90)
弓正形贝科 Toxorthidae Rong,1984	(90)
微小正形贝属(新属) *Minutorthis* Zeng,Chen et Zhang(gen. nov.)	(90)
五房贝目 Pentamerida Schuchert et Cooper,1931	(91)
五房贝亚目 Pentameridina Schuchert et Cooper,1931	(91)
五房贝超科 Pentameroidea M'Coy,1844	(91)
五房贝科 Pentameridae M'Coy,1844	(91)
槽五房贝属 *Sulcipentamerus* Zeng,1987	(91)
中褶贝属(新属) *Centreplicatus* Zeng,Zhang et Han(gen. nov.)	(93)
从五房贝属 *Apopentamerus* Boucot et Johnson,1979	(94)
围板贝亚属(新亚属) *Apopentamerus*(*Enclosurus*) Zeng,Wang et Peng (subgen. nov.)	(95)

参考文献 ...... (97)

Abstract ...... (101)

索引 ...... (109)

图版和图版说明 ...... (113)

Contents

Chapter 1　Strata, Age and preservable form of fossils ……………………………… (1)
  1.1　Strata ……………………………………………………………………………………… (1)
  1.2　Age ………………………………………………………………………………………… (3)
  1.3　Preservable form of fossils …………………………………………………………… (3)

Chapter 2　Turbidite deposit and incompleteness of the Guanyinqiao Member ………… (4)
  2.1　Turbidite ………………………………………………………………………………… (4)
  2.2　Incompleteness of the Guanyinqiao Member(Bed) ………………………………… (6)

Chapter 3　Habitat of *Hirnantia* Fauna and geological events ……………………… (7)
  3.1　Geological events ……………………………………………………………………… (7)
  3.2　Ecologic niche and ecotype during early Late Katian ……………………………… (7)
  3.3　Ecologic niche and ecotype from late Late Katian to Early Hirnantian …………… (9)
  3.4　Palaeoecologic environment of *Hirnantia* Fauna …………………………………… (9)

Chapter 4　Brief discussion on extinctive cause of the *Hirnantia* Fauna ………… (11)
  4.1　Extinctive time …………………………………………………………………………… (11)
  4.2　Marker and ecologic niche …………………………………………………………… (11)
  4.3　Extinctive cause of the *Hirnantia* Fauna …………………………………………… (11)

Chapter 5　Biotic evolution in Llandoverian in Yichang District …………………… (13)
  5.1　Fauna of Longmaxian …………………………………………………………………… (13)
    5.1.1　Environment and fauna of Early Longmaxian ………………………………… (13)
    5.1.2　Environment and fauna of Late Longmaxian …………………………………… (13)
  5.2　Great resuscitations of Biota during Luorepinian …………………………………… (15)
    5.2.1　Environments of Luorepinian …………………………………………………… (15)
    5.2.2　Great resuscitations of Biota during Luorepinian ……………………………… (15)
  5.3　The Upper Yangtze Sea towards doom ………………………………………………… (16)
    5.3.1　Sediments and contact relations ………………………………………………… (16)
    5.3.2　Evidences of the Upper Yangtze Sea towards doom …………………………… (17)

Chapter 6　Systematic description of fossils ………………………………………… (18)
  6.1　Fossils of *Hirnantia* Fauna …………………………………………………………… (18)
  Phylum Brachiopoda ………………………………………………………………………… (18)
  Subphylum Linguliformea …………………………………………………………………… (18)
    Class Lingulata ………………………………………………………………………… (18)
      Order Lingulida …………………………………………………………………… (18)
        Superfamily Discinoidea ……………………………………………………… (18)
          Family Discinidae ……………………………………………………… (18)

Genus *Orbiculoidea*	(18)
Subphylum Craniiformea	(19)
Class Craniata	(19)
Order Craniopsida	(19)
Superfamily Craniopsoidea	(19)
Family Craniopsidae	(19)
Genus *Craniops*	(19)
Order Craniida	(20)
Superfamily Cranioidea	(20)
Family Craniidae	(20)
Genus *Acanthocrania*	(20)
Genus *Philhedra*	(21)
Subphylum Rhynchonelliformea	(21)
Class Strophomenata	(21)
Order Strophomenida	(21)
Superfamily Strophomenoidea	(21)
Family Strophomenidae	(21)
Subfamily Furcitellinae	(21)
Genus *Minutomena*(gen. nov.)	(21)
Family Rafinesquinidae	(23)
Subfamily Leptaeninae	(23)
Genus *Leptaena*	(23)
Genus *Leptaenopoma*	(25)
Family Glyptomenidae	(29)
Subfamily Glyptomeninae	(29)
Genus *Paromalomena*	(29)
Family Sinomenidae(fam. nov.)	(30)
Genus *Aphanomena*	(31)
Genus *Eostropheodonta*	(32)
Genus *Yichangomena*(gen. nov.)	(34)
Genus *Sinomena*(gen. nov.)	(36)
Genus *Hubeinomena*(gen. nov.)	(37)
Superfamily Plectambonitoidea	(38)
Family Xenambonitidae	(38)
Subfamily Aegiromeninae	(38)
Genus *Aegiromena*	(38)
Genus *Aegiria*	(41)
Genus *Chonetoidea*	(43)
Genus *Trimena*(gen. nov.)	(45)
Order Orthotetida	(46)
Suborder Orthotetidina	(46)
Superfamily Orthotetoidea	(46)
Family Fardeniidae	(46)

- Subfamily Fardeniinae ……………………………………………………………… (46)
  - Genus *Fardenia* ……………………………………………………………… (46)
  - Genus *Coolinia* ……………………………………………………………… (47)
- Suborder Triplesiidina ……………………………………………………………… (48)
  - Superfamily Triplesioidea ……………………………………………………… (48)
    - Family Triplesiidae …………………………………………………………… (48)
      - Genus *Triplesia* …………………………………………………………… (48)
      - Genus *Cliftonia* …………………………………………………………… (50)
- Class Rhynchonellata ……………………………………………………………… (53)
  - Order Orthida ……………………………………………………………………… (53)
    - Sudorder Orthidina ……………………………………………………………… (53)
      - Superfamily Plectorthoidea …………………………………………………… (53)
        - Family Toxorthidae ………………………………………………………… (53)
          - Genus *Toxorthis* ………………………………………………………… (53)
  - Suborder Dalmanellidina ………………………………………………………… (54)
    - Superfamily Dalmanelloidea …………………………………………………… (54)
      - Family Dalmanellidae ………………………………………………………… (54)
        - Subfamily Dalmanellinae …………………………………………………… (54)
          - Genus *Dalmanella* ……………………………………………………… (54)
          - Genus *Onniella* ………………………………………………………… (56)
          - Genus *Trucizetina* ……………………………………………………… (58)
          - Genus *Mirorthis* ………………………………………………………… (61)
          - Genus *Paramirorthis* (gen. nov.) ……………………………………… (62)
    - Superfamily Enteletoidea ……………………………………………………… (64)
      - Family Draboviidae …………………………………………………………… (64)
        - Subfamily Draboviinae ……………………………………………………… (64)
          - Genus *Drabovia* ………………………………………………………… (64)
          - Genus *Drabovinella* …………………………………………………… (66)
          - Genus *Hirnantia* ………………………………………………………… (68)
          - Genus *Kinnella* ………………………………………………………… (74)
          - Genus *Draborthis* ……………………………………………………… (77)
      - Family Dysprosorthidae (fam. nov.) ………………………………………… (80)
        - Genus *Dysprosorthis* ……………………………………………………… (80)
  - Order Rhynchonellida ……………………………………………………………… (82)
    - Superfamily Rhynchotrematoidea ……………………………………………… (82)
      - Family Trigonirhynchiidae …………………………………………………… (82)
        - Subfamily Rostricellulinae ………………………………………………… (82)
          - Genus *Plectothyrella* …………………………………………………… (82)
  - Order Athyridida …………………………………………………………………… (84)
    - Suborder Athyrididina …………………………………………………………… (84)
      - Superfamily Meristelloidea …………………………………………………… (84)
        - Family Meristellidae ………………………………………………………… (84)
          - Subfamily Meristellinae …………………………………………………… (84)

    Genus *Hindella* .... (84)
6.2 Some new materials of brachiopods in Luorepinian .... (88)
Order Strophomenida .... (88)
 Superfamily Plectambonitoidea .... (88)
  Family Leptellinidae .... (88)
   Subfamily Leptellininae .... (88)
    Genus *Leptellina* .... (88)
     Subgenus *Leptellina(Merciella)* .... (88)
    Genus *Leptelloidea* .... (89)
Order Orthida .... (90)
 Suborder Orthidina .... (90)
  Superfamily Plectorthoidea .... (90)
   Family Toxorthidae .... (90)
   Genus *Minutorthis*(gen. nov.) .... (90)
Order Pentamerida .... (91)
 Suborder Pentameridina .... (91)
  Superfamily Pentameroidea .... (91)
   Family Pentameridae .... (91)
   Genus *Sulcipentamerus* .... (91)
   Genus *Centreplicatus*(gen. nov.) .... (93)
   Genus *Apopentamerus* .... (94)
   Subgenus *Apopentamerus(Enclosurus)* (subgen. nov.) .... (95)

References .... (97)

Abstract .... (101)

Index .... (109)

Plates and explanations .... (113)

第1章　地层和时代以及化石保存形式

1.1　地层

目前所指的地层仅为五峰组观音桥段，厚17～30cm。

宜昌地区五峰组观音桥段（或称观音桥层）广泛出露于黄陵背斜东、西两翼，其中以东翼的岩层产状相对较缓，出露较好，化石极为丰富，研究精度较高。目前的化石材料就是采集于黄陵背斜东翼、宜昌市以北沿着去保康公路22～42km之间的黄花场、丁家坡、王家湾3条剖面五峰组观音桥段（插图1）。宜昌地区观音桥段其原岩应为泥灰岩或钙质泥岩，这在王家湾小河东岸边，以及在王家湾以北不远处的保康县梨树垭小山丘等地方仍然保存有较好的泥灰岩就是例证。但是多数地方由于遭到强烈风化或遭受到沉积龙马溪组黑色页岩段时偏弱酸性海水长时间浸泡、其钙质被融蚀掉的缘故，因此在野外所见的几乎都为水云母黏土岩，并可分为3部分：下部为黑灰、黄褐色或浅紫灰色水云母黏土；但这部分地层由于曾经遭到当时海底浊流侵蚀，已不太完整，甚至在个别地方已全部缺失（详见第2章浊流章节）。中部为黄灰色、米黄色含石英水云母黏土岩，其中石英粒径0.01～0.99mm，但以0.1～0.25mm为多，分选不好，为次棱角状到半圆状（图版62，图1），部分石英颗粒呈碎玻璃状，个别的有火山玻璃质包体，为火

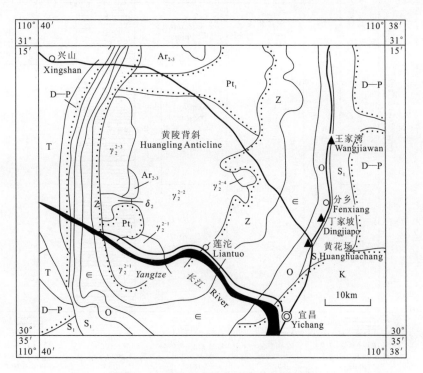

插图1　宜昌地区交通位置和简化地质图

Illustr. 1　Location and simplified geological map of Yichang District, Western Hubei, China

▲剖面位置（sections locations）

山晶屑(为流纹质？层凝灰岩)(曾庆銮等,1983,1987)。上部为黑灰色、黄褐色、浅灰色含石英粉砂质水云母黏土岩。宜昌地区观音桥段上、中、下3部分都产极为丰富的赫南特腕足动物群(Hirnantia Fauna)的化石。

观音桥段上覆地层为龙马溪组黑色页岩段,两者呈整合接触,岩性为黑色或黑灰色薄层泥质或硅质页岩,产极为丰富的 Normalograptus persculptus 笔石带的笔石化石群。观音桥段下伏地层为五峰组笔石页岩段,彼此间呈整合接触,岩性为黑色或黑灰色薄层泥质或硅质页岩,其上部产极为丰富的 Normalograptus extraordinarius 笔石带的笔石化石群(插图2)。

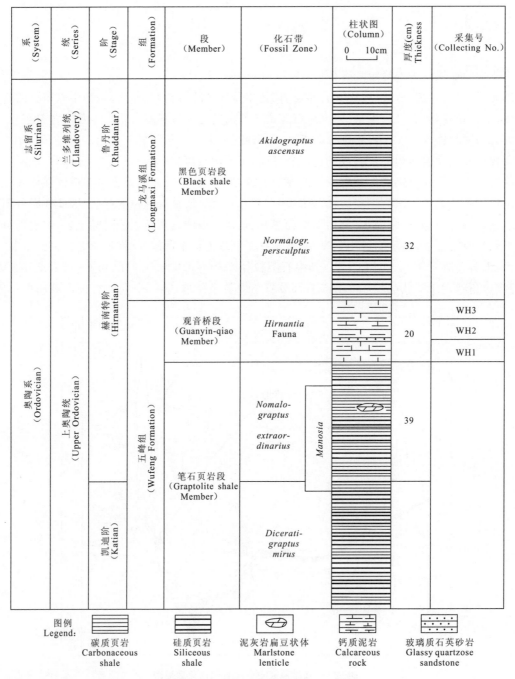

插图 2 王家湾上奥陶统赫南特阶柱状图

Illustr. 2 The columar section of Hirantian Stage of the Upper Ordovician Series in Wangjiawan, Yichang District

1.2 时代

宜昌王家湾剖面是上奥陶统顶部赫南特阶（Hirnantian）全球层型剖面和点位建立的所在剖面（陈旭，戎嘉余等，2006）。而五峰组观音桥段（层）则位于赫南特阶的中部，也是赫南特动物群（Hirnantia Fauna）产出层位；赫南特阶的下部则为五峰组笔石页岩段最上部 *Normalograptus extraordinarius* 带的所在层位；赫南特阶的上部，便是龙马溪组黑色页岩段最下部 *Normalograptus persculptus* 带的产出层位（插图2）。

因此观音桥段的地质时代为晚奥陶世赫南特中期（Middle Hirnantian of Late Ordovician），即介于 *Normalograptus extraordinarius* 带与 *Normalogr. persculptus* 带之间。

1.3 化石保存形式

从宜昌地区观音桥段内所采集到的赫南特动物群的化石几乎都是以内模的形式保存，哪怕是在2m深探槽里、很新鲜的岩石内的化石也是如此。究其原因，很可能是宜昌地区在沉积龙马溪组黑色页岩段时全球正在发生大海侵，而这一大海侵是由于冈瓦纳古陆大冰期融化所导致的（Sutcliffe et al.，2001）。随着冰盖消融速度的加快，海侵也在逐步加大。由于海水逐步加深，当时的表层水又布满着浮游笔石动物群，耗尽表层水的含氧量，造成当时的水介质一是严重缺氧，二是 CO_2 不能逸出，加上大量笔石遗体在海底的腐烂作用所产生的 CO_2，更是导致当时的水介质充满着 CO_2。在此情况下，一般在水深500m以下就能致使 $CaCO_3$ 长期处在严重不饱和的状态而不能沉淀（王英华，1979）。这也就是说当时的海水是属于偏酸性的。而宜昌地区观音桥段仅有17～30cm厚，长期浸泡在沉积龙马溪组黑色页岩段时偏酸性的海水之下，其中所含的赫南特动物群的钙质壳及其钙质骨骼必然会全部被溶蚀掉。这也许是造成赫南特动物群的化石都是以内模形式保存的根本原因。

第 2 章 浊流沉积及观音桥段的不完整性

2.1 浊流沉积

著者之一(曾庆銮)在进行劈开观音桥段大块岩石、精选化石的过程中,发现在观音桥段中部(即化石采集号 WH2)之内存在着侵蚀面、碎屑岩层和介壳化石碎片层,心里感到迷惑不解,在较深水域(为广海外陆棚区)沉积的岩层中怎么会存在着这种现象。于是去请教沉积岩学家杨振强教授,他一看就识别出此现象是深水区产生浊流沉积的结果。因此我们对宜昌地区观音桥段中的浊流沉积进行研究,并选择宜昌地区观音桥段出露较好的黄花场、丁家坡、王家湾 3 个剖面进行野外观察。结果发现这 3 个剖面都存在着浊流沉积层序。其情况是:发生浊流的时间是在奥陶纪末期的 Hirnantian 中期,产生的具体层位为观音桥段中部(即本书的化石采集号 WH2 的中部)。此 3 个剖面的浊流沉积层序大同小异,只是浊流侵蚀强度有高有低,侵蚀的槽模有存在于不同层位上的现象。现将比较典型的王家湾剖面(图版 61,图 2)和丁家坡剖面(图版 61,图 1)的浊流沉积层序分别描述如下。

2.1.1 王家湾观音桥段浊流沉积剖面

王家湾观音桥段浊流沉积层序(图版 61,图 2)从上到下描述如下:

A. 为正常环境化石埋藏层,岩性为生物钙质泥岩,产极为丰富的腕足类化石,介形虫、三叶虫和腹足类化石稀少。

B. 以腕足类化石碎片为主,含少量小型海百合茎、三叶虫和腹足类等化石碎片,岩性为生物碎片钙质泥岩(图版 62,图 2),厚 3～5mm。

C. 为由下到上从粗到细的粒度递变层(gradebed)。其厚度变化大,横向变化快,碎屑颗粒大小不均,分选很差,呈次棱角状至半圆状。底部以大颗粒为主,并且主要集中在凹坑状的槽模内,碎屑成分以下伏岩层的岩性和长石为主,含少量石英及生物碎片或生物屑(图版 61,图 2;插图 3),厚 6～19mm。

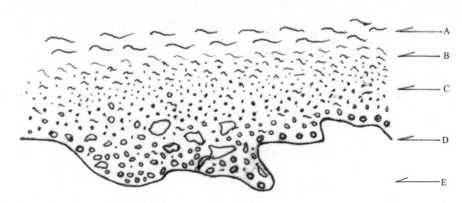

插图 3 浊流沉积层序示意图(依据图版 61,图 2)

Illustr. 3 A sketch map presenting the sedimentary sequences of turbidite deposite in the study area(from pl. 61,fig. 2)

A-正常环境化石埋藏层;B-介壳碎片层;C-粒度递变层;D-浊流侵蚀 WH2 地层段的槽模;E-浊流沉积层序的基层

粒度递变层中部以石英砂粒为主(图版62,图1)含少量长石和生物碎片。石英粒径0.01~0.99mm,但以0.1~0.25mm为多,分选差,为次棱角状至半圆状。部分石英颗粒呈碎玻璃状,个别的有火山玻璃质包体,为火山晶屑。其实这层石英在20世纪80年代初,汪啸风等进行长江三峡地区生物地层学(2)研究的时候就已经发现,当时的岩石薄片由徐安武教授鉴定,认为此岩性与火山活动有关,并定为流纹质(?)层凝灰岩(见曾庆銮等,1987)。厚3~11mm。

粒度递变层上部为含少量细石英砂和泥灰岩屑及少量生物碎片钙质泥岩。厚10~14mm。

D. 为浊流(湍流-turbulent)运动时侵蚀浊流之前的沉积层,并挖成的槽模(flute casts)。

E. 为浊流沉积层序的基层,即观音桥段中部(化石采集号WH2的地层段)。

2.1.2 丁家坡观音桥段浊流沉积剖面

宜昌丁家坡观音桥段浊流沉积层序(图版61,图1)从上到下分述如下:

A. 为正常环境化石埋藏层,岩性为生物泥灰岩或钙质生物泥岩,产极为丰富的腕足类化石,少量三叶虫、介形虫和腹足类。

B. 含大量腕足类碎片钙质泥岩,厚3~5mm,情况与王家湾剖面差不多。

C. 为由下到上从粗到细的粒度递变层。岩屑主要为长石、钙质泥岩、石英、黑色硅质岩以及少量腕足类碎片和小型海百合茎。底部岩屑颗粒相对较粗,越往上颗粒越细。分选都不好,磨圆度也较差。本层与王家湾剖面相比,岩屑颗粒和石英粒径明显变小,多数变为石英细砂泥岩或石英粉砂泥岩。厚4~7mm。

D. 为浊流(湍流)运动时在五峰组笔石页岩段最顶层的黑色硅质岩表面侵蚀所挖出来的槽模。

E. 以五峰组笔石页岩段最顶部的一层黑色硅质岩(厚4~6cm)为本浊流沉积层序的基层。

2.1.3 浊流依据与成因

2.1.3.1 浊流的依据

(1)宜昌地区在Hirnantian时期的观音桥段是夹在五峰组笔石页岩段与龙马溪组黑色笔石页岩段之间,并处在上扬子海外陆棚区,即BA3—BA4底栖组合的一种生态位(曾庆銮,1991)。徐安武(1987)对五峰组观音桥段的岩石薄片进行鉴定,并对其沉积相进行分析,认为是广海陆棚相,也证明当时的水体是比较深的。因此在沉积观音桥中部的时候,当时宜昌地区是属于汪洋大海,绝对不可能暴露于地表而产生那样的剥蚀面(槽模或凹坑),那样的侵蚀槽模只有在较深海底产生浊流时才能形成,并且得到保存,而不遭到后期的改造和破坏(许靖华,1980)。因此槽模是浊流最重要的依据。

(2)在槽模内沉积着从下到上、由粗到细的粒度递变层。其岩屑多数来自于下伏岩层,分选性和磨圆度差。上述这些特点都是由于当时海底突然产生急流,对海底上部的海泥层进行冲刷、侵蚀,并将许多冲碎的岩屑和泥浆一起带走,从而形成浊流。当流速减慢粗粒屑先沉下,细粒屑在后,有更大的岩屑只进行短距离的位移,而生物碎片漂浮力较强,沉积在上面,因而形成如王家湾剖面那样的粒度递变层(图版61,图2),而粒度递变层则是浊流沉积最为重要的证据之一。

(3)浊流沉积的保存条件。海底里的浊流沉积层序只有在水体较深、海浪影响不到,而又较为安宁的海域里才能保存下来,否则将被改造掉。而宜昌地区当时沉积观音桥段时的生态底栖组合位置为BA3—BA4(或者为BA4),为外陆棚区内侧,位于浪基面以外的海域,正符合保存浊流沉积层序的条件。因此王家湾剖面的浊流沉积层序才会保存得那么完整(图版61,图2)。因此在深水域有那样的沉积特征,也可以证明它是浊流的产物。

2.1.3.2 浊流的成因

在较为深水而又较为安宁的海底要突然产生一个侵蚀作用很强的水下流,从而形成浊流,这首先要有产生急流的动能。粒度递变层中含有大量石英颗粒尤为重要,从曾庆銮(1987)宜昌黄花场奥陶系剖面描述的第 53 层中得知,那些石英的粒径为 0.01~0.99mm,但以 0.1~0.25mm 为多,分选不好,为次棱角状至半圆形;部分石英颗粒呈碎玻璃状,个别的有火山玻璃质包体,为火山晶屑。含有此石英颗粒的岩石被定为流纹质(?)层凝灰岩(徐安武,1987)。因此推测,当时突然产生侵蚀作用很强的水下湍流的动能可能是由当时在宜昌地区附近曾发生过短暂的海底火山喷发所引起,导致当时海底产生急流,并且强烈侵蚀之前的海底泥,挖出许多槽模,从而形成前述的浊流沉积层序。

2.2 观音桥段的不完整性

从前面对宜昌王家湾浊流沉积层序剖面和宜昌丁家坡浊流沉积层序剖面的描述得知:王家湾剖面 D 层的侵蚀槽模是产生在观音桥段中部,即 WH2 层位的中部(图版 61,图 2)。而丁家坡剖面 D 层的侵蚀槽模是发生在五峰组笔石页岩段最顶层的黑色硅质页岩上(图版 61,图 1)。这就是说在丁家坡剖面相当于王家湾剖面的 WH2 下部和相当于 WH1 的所有地层(共厚约 13~14cm)已被浊流侵蚀掉。因此当时所产生的浊流在宜昌地区的观音桥段至少要被侵蚀掉 13~15cm 厚的地层。说明现在宜昌地区观音桥段各个剖面的厚度不是当时沉积应有的厚度,而是或多或少都受到当时所产生的浊流侵蚀的影响。

第 3 章　赫南特动物群的生境和地质事件

3.1　地质事件

奥陶纪凯迪期晚期至志留纪鲁丹期早期这段地史时期在整个地史长河中虽然是很短暂的,但是却发生了一连串非常引人注目而又重大的地质事件:首先是在 *Tangyagraptus typica* 带笔石生存期间,全球发生过一次罕见的大海侵事件;其二是 *Diceratograptus mirus* 笔石带至 *Normalograptus extraordinarius* 笔石带生存期间,开始在以非洲北部为中心,后来扩展到整个冈瓦纳古大陆的大冰期地质事件;其三是由于冈瓦纳古大陆冰盖向大陆架进一步扩大,造成全球大海退事件;其四是由于全球大海退事件,导致海平面强烈下降,致使喜欢生存于深水区而又广泛分布于世界许多地区,相当于五峰期那样的笔石动物群的大灭绝事件;其五是由于该大海退,致使当时宜昌地区所在的上扬子海变为底栖组合 4(BA4)的一种生态位,成为著名赫南特动物群非常优良的繁殖地,繁衍了 15 个科、35 个属、43 个种这个庞大的赫南特动物群;其六是随着冈瓦纳古陆大冰盖的融化,导致从赫南特晚期至鲁丹早期的大海侵事件,致使世界著名赫南特动物群的大灭绝。

为了能够更好地反映上述重大事件的演变过程,以及其对当时宜昌地区所在上扬子海的影响,并能更好地对其进行讨论。我们采用了 Zeigler(1965)把陆棚区从近岸到远岸依次划分为 5 个腕足动物群落栖息带,并把陆棚区以外的海域划为以笔石为主的浮游群落领域的划分方法;同时以 Boucot(1975)的 6 个底栖组合 1~6(Benthic Assemblags 1~6,即 BA1—BA6)的划分理念,以及戎嘉余(1986)的 5 个底栖组合带(BA1—BA5)在陆棚区相应生态位置的模式,将宜昌地区从 Late Katian 到 Early Rhuddanian 期间的古生态环境划分为 5 个演变阶段(插图 4)。

3.2　晚凯迪早期的生态底域位置及其生态类型

3.2.1　发生时间

当时宜昌地区的上扬子海是处在海侵高峰,其生态底域位置为底栖组合 6(BA6)。发生时间为晚凯迪早期,即 *Tangyagraptus typicus* 笔石带生存期间(插图 4-A)。

3.2.2　标志物和生态底域位置

当时上扬子海保存下来最为重要的物证有三:①黑色或黑灰色薄层硅质页岩;②在黑色硅质页岩中堆积着大量的笔石化石;③在黑色硅质页岩内含有大量的放射虫化石(徐安武,1987,图版 68,图 1;曾庆銮,1987;曾庆銮,1991,图版 2,图 4)。上述 3 种海洋沉积物都被公认为远洋深水(BA6)最为重要的标志物(Ziegler,1965;Boucot,1975;郑宁等,2012)。因此推断宜昌地区在 *Tangyagraptus typicus* 笔石带生存期间的上扬子海的生态底域位置处在底栖组合 6(BA6)。

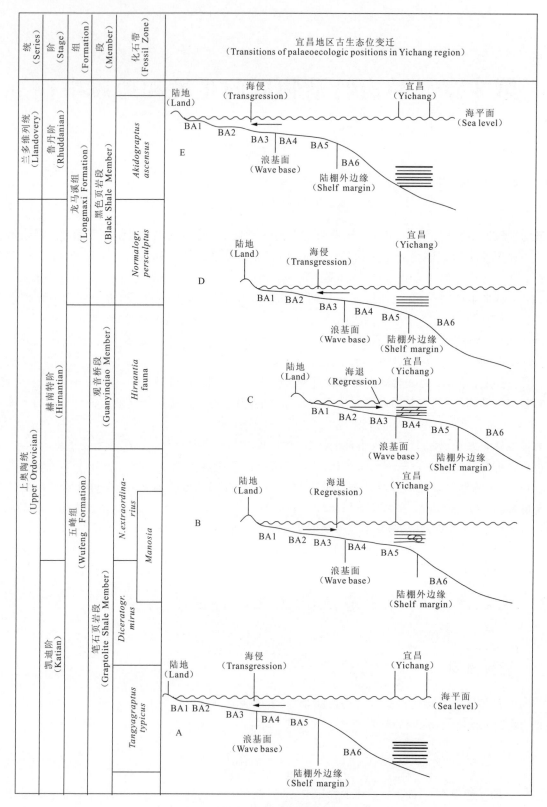

插图 4　宜昌地区晚凯迪期至早鲁丹期生态位变迁示意图

Illustr. 4　A sketch map showing transitions of the palaeoecologic positions from late Katian to early Rhuddanian in Yichang region

据上述标志物推断:一是当时宜昌地区所在的上扬子海的整个表层水是布满着浮游笔石动物群和放射虫动物群;二是当时的海水严重滞流,海水底层严重缺氧;三是当时的海水中 $CaCO_3$ 严重不饱和,始终未能沉淀且不含钙质,底栖动物(钙质壳类)无法生存。

3.3 晚凯迪晚期至早赫南特期的生态底域位置及其生态类型

3.3.1 发生时间

当时宜昌地区所在的上扬子海开始发生幅度不太大的海退,发生时间为 *Diceratograptus mirus* 笔石带至 *Normalograptus extraordinarius* 笔石带生存期间(插图 4-B)。

3.3.2 标志物和生态底域位置

当时宜昌地区所在的上扬子海保存下来最为珍贵的标志物有三:①含有稀少,并以 *Manosia* 为代表的钙质壳腕足动物群落化石;②黑灰色页岩中偶尔含有黑灰色泥灰岩扁豆状体;③黑灰色薄层笔石页岩偶尔夹有薄层硅质岩。

从上述3种标志物的性质来看,明显反映当时上扬子海的海平面比其前期有所下降,变成 *Manosia* 群落有时能够勉强生存,能够从海水中吸收一些钙质去营造它们的贝体。但当时的海水仍然较深, $CaCO_3$ 不能经常达到饱和,海水中钙质较为稀缺,满足不了大量腕足生存的需要。另一方面,当时的笔石化石还极为丰富,并伴有少量黑色薄层硅质页岩存在,表明当时的海水还较深。因而推断当时宜昌地区所在的上扬子海是处在底栖组合 5-6(BA5—BA6)的一种生态底域位置(插图 4-B)。而这一不太明显的海退事件明显与冈瓦纳古陆开始发生冰盖事件紧密相关(戎嘉余,1984),而且应是受到早赫南特期冈瓦纳古陆大冰盖成长期(Sutcliffe et al.,2001)的影响所导致的。

3.3.3 古生态环境

当时宜昌地区所在的上扬子海与其前期深水区的环境有所不同,变成在其海底有时能够勉强生存着钙质壳腕足类 *Manosia* 群落,表明当时的海水有质上的变化。但当时海水仍然较深,氧含量还严重不足,水中的钙质还是较为稀缺,满足不了大量腕足动物生存的需要,仍然显得更适应于浮游笔石动物群生存的一种生态环境。

3.4 赫南特动物群的生态环境

3.4.1 赫南特动物群生存的时间

宜昌地区赫南特动物群生存在著名的冈瓦纳古陆冰盖向陆架区大为扩张(expansion of the ice sheet onto shelf)期间,即介于 *Normalograptus extraordinarius* 笔石带与 *Normalograptus persculptus* 笔石带之间,当时的上扬子海正处在海退时期(插图 4-C)。因此宜昌地区赫南特动物群应是冈瓦纳古陆大冰盖顶峰期的产物。

3.4.2 标志物及生态底域位置

冈瓦纳古陆冰盖向陆架区大为扩张期间(Sutcliffe et al. 2001,fig.7-C),导致全球大海退,当时的上扬子海也跟随着大海退,并沉积和保留下来最为珍贵的物证有三:①极为丰富、分异度高的 *Hirnantia* Fauna 化石群,并与少量 *Dalmanitina* 为代表的三叶虫群,以及 *Bairdia* 为代表的介形虫群等化石共存;②黄灰色或浅紫灰色钙质泥岩或薄层泥灰岩;③保存完好的浊流沉积层序。

宜昌地区赫南特动物群生存的生态底域位置,作者之一(曾庆銮等,1991,2015b)曾确定为底栖组合3-4(BA3—BA4),现在我们经过重新鉴别,其生态底域位置应属于 BA4 更为合适。其理由:①宜昌地区赫南特动物群大多数成员的贝体都很小(壳宽 2～5mm 的占多数,参见本书化石系统描述),即使较大的贝体,但都是很扁平,仍然是较适应于较深水区的海底软泥中生活;②目前我们在宜昌地区观音桥段中发现保存完好的浊流沉积层序,而此种沉积层序只有在较深水域下才能保存下来,否则会被后期的海浪改造掉(许靖华,1980)。依据上述原因,我们将宜昌地区观音桥段赫南特动物群的生态底域位置改为底栖组合4(BA4)。

3.4.3 两大地质事件及赫南特动物群的生态环境

3.4.3.1 两大地质事件

由于冈瓦纳古陆大冰盖向陆架区大为扩张,导致全球性大海退,致使海平面急剧下降,造成全球性生态环境大变化,其结果是:

(1)导致喜欢生存于深水区、严重缺乏 $CaCO_3$ 的那种水质环境中的笔石动物群(相当于五峰组笔石页岩段的笔石动物群)大灭绝事件。

(2)由于该全球性的大海退对喜欢在深水区生存的浮游动物群来说是一种灭顶之灾,但对于喜欢在陆架区生存的底栖动物而言却是一种天大的良机,因而产生了广泛分布于世界各地的赫南特动物群事件。

3.4.3.2 宜昌地区赫南特动物群的生存环境

从前述的情况得知,当时宜昌地区所在的上扬子海是处于底栖组合 4(BA4)的一种生态底域位置。其海水深度适中,很安宁,海水中 $CaCO_3$ 的饱和程度也很合适(岩性为泥灰岩或钙质泥岩),完全能够满足腕足动物吸收去营造它们的贝壳。另外,当时宜昌地区所在的上扬子海远离冈瓦纳古陆冰盖区,其水温相对较高,甚至有可能接近于暖水型。这从与宜昌地区相距不太远、同样属于上扬子海的贵州沿河观音桥段产有珊瑚类 *Amsassia* sp. 和四川綦江观音桥段产有珊瑚类 *Protoheliolite* sp. 便可以引证。因为珊瑚类也是营底栖固着动物,喜欢生存在暖水(约 23～25℃)、含盐度正常(35‰)、水介质中含有足够碳酸钙的陆棚区内。因此,当时宜昌地区赫南特动物群生存时的海水应接近于暖水型或为温水型,这与前人所指的 *Hirnantia* Fauna 为冷水型的(Temple,1965;Havliček,1989)有所不同。可能也正因为如此,当时宜昌地区的上扬子海才成为赫南特动物群得天独厚、大量繁衍的一种海域。从目前的资料得知,当时宜昌地区的上扬子海几乎集聚了世界各地赫南特动物群所有成员(达 15 个科、35 个属、43 个种)的昌盛景象。

第 4 章 赫南特动物群灭绝原因的探讨

4.1 灭绝时间

赫南特动物群(Hirnantia Fauna)灭绝于晚赫南特早期,即 *Normalograptus persculptus* 笔石带生存期间的最早期。

4.2 标志物和生态底域位置

4.2.1 标志物

(1)龙马溪组黑色页岩段底部的黑色薄层硅质笔石页岩中产极为丰富的笔石化石。
(2)五峰组观音桥段顶部黄灰色钙质泥岩中产极为丰富的腕足类化石。
(3)观音桥段与龙马溪组黑色页岩段之间的转换分界面上见有少许赫南特动物群成员的化石与大量 *Normalograptus persculptus* 带的笔石群化石共存(图版 63)。

4.2.2 生态底域位置变迁

(1)从前述的第 3 章得知,宜昌地区观音桥段的生态底域位置是处在底栖组合 4(BA4)。
(2)根据龙马溪组黑色页岩段产极为丰富的笔石化石推断,当时宜昌地区的上扬子海开始发生大海侵,其生态底域位置是处在底栖组合 6(BA6)。而这一大海侵,又与当时冈瓦纳古陆大冰盖消融、造成全球大海侵事件(Sutcliffe et al.,2001)紧密相关。以上说明,当时宜昌地区的上扬子海从中赫南特期(Middle Hirnantian),即 *Hirnantia* Fauna 生存期间的生态底域位置是处在 BA4,而到晚赫南特期(Late Hirnantian),即 *Normalograptus persculptus* 笔石带生存期间的生态底域位置变为 BA6(插图 4-D,E)。由于生态位的变迁,必然导致生存在该地区动物群的更替,从适合于底栖腕足类的生活环境转换为符合于浮游笔石动物群的生态环境。

4.3 赫南特动物群灭绝的原因

据目前所知,造成赫南特动物群灭顶之灾的原因有下列 4 种见解:
(1)由伽马(γ)射线(Gamma ray)辐射造成的。
(2)由于冈瓦纳古陆冰盖向陆架区扩张,导致全球性海平面大幅度下降所造成的。
(3)由于海底火山爆发释放毒气造成的。
(4)天外铱大量降落,导致铱异常造成的。

我们对上述4种见解不想妄加评论，但也不能苟同。

我们认为造成赫南特动物群灭顶之灾的罪魁祸首是浮游笔石动物群。这不是由于浮游笔石动物本身凶猛地去捕杀别的动物群，而是由于笔石动物的生活习性所决定的。浮游笔石动物群都是以浮游在表层水的生活方式为生。当它们迅猛发展到布满整个表层水时，必然耗尽表层水的含氧量（全被笔石动物群吸收），同时遮住阳光的照射。加上大量笔石遗体下沉于海底形成腐烂作用，CO_2充满于海底，这样必定导致下层水严重缺氧，致使水质极差，从而毒死底栖动物。这犹如近代海洋、湖泊、鱼塘，只要是蓝藻、浮萍、水葫芦等任何一种的生物在其表层水泛滥，既挡住阳光照射又耗尽表层水的含氧量，其水下的任何动物都会遭到灭顶之灾，都会被毒死是同一个道理的。

目前根据龙马溪组黑色页岩段产极为丰富甚至呈堆积状态的笔石化石来推断，当时的浮游笔石动物群必定是布满于整个表层水，它们既挡住阳光的照射，又耗尽整个表层水的含氧量，造成底层水严重缺氧，水质极差，从而毒杀赫南特动物群。这就是造成距今约444.33Ma前，并且只经历约63万年（0.63Myr±）历程的赫南特动物群灭绝的根本原因。

第 5 章　宜昌地区兰多维列世生物群演变

5.1 沉积龙马溪组时的动物群

5.1.1 早期古生态环境和动物群

沉积龙马溪组早期时(相当于沉积龙马溪组黑色页岩段时),由于冈瓦纳(Gondwana)古陆大冰期发生消融事件(Sutcliffe et al.,2001),导致全球性发生大海侵,同时也造成宜昌地区在兰多维列世早期(Early Llandoverian)发生兰多维列世第一幕大海侵,并引来冷水型的海水。因此当时宜昌地区的生态底域位置演变成为 BA6(插图 5-A),并沉积一套 51.9m 厚的龙马溪组黑页岩段(汪啸风等,1987)。在此期间,宜昌地区是处在水体深度大(BA6),导致水介质严重缺氧、海底充满 CO_2(因为水体深 CO_2 不能逸出,加上当时海底域沉积着大量笔石遗体产生腐烂作用所形成的 CO_2),导致 $CaCO_3$ 长期不能沉淀的一种海域(曾庆銮等,2015b)。因此当时宜昌地区所处的海域只有在表层水独自生存着极为昌盛的,并且以 *persculptus* 带、*ascensus*(或 *acuminatus*)带、*atavus* 带、*acinaces* 带、*cyphus* 带、*triangulatus* 带、*magnus* 带、*argenteus* 带等笔石带为代表的浮游笔石动物群,而任何底栖动物都无法在此海域生存的一种古生态环境。这从其沉积物为黑色硅质页岩或黑色页岩中仅含有极为丰富的笔石化石也可以为以上的结论提供有力的佐证。

5.1.2 晚期古生态环境和动物群

宜昌地区在沉积龙马溪组黑色页岩段之后,便发生宜昌地区兰多维列世(Llandoverian)第一幕大海退,当时快速沉积一套 571m 厚的细碎屑岩(粉砂质页岩与粉砂岩或细砂岩互层)的黄绿色页岩段(汪啸风等,1987)。本段下部产有少量笔石化石,但由于越往上砂质越多,海水越浅,笔石化石越少,有许多地层段很难发现,甚至没有。直到黄绿色页岩段上部偏下(相当于 *arcuata* 带中部)还产有"*Lingula*"sp.(汪啸风等,1987)。而"*Lingula*"这一类型的腕足类则为生态底域位置 BA1 的典型代表(Ziegler,1965;Boucot,1975;戎嘉余,1986;戎嘉余等,1984;曾庆銮,1991;曾庆銮等,2015b),由此说明当时宜昌地区的生态位已退到 BA1 的一种海域(插图 5-C)。而这一大海退事件在上扬子区的贵州北部和东北部、川南等地表现得更为明显,有不少地方发现有风化壳和泥裂等现象(戎嘉余等,2012)。戎嘉余等将黔北这一地区的海退事件称为"桐梓上升"。而发生"桐梓上升"地质事件的时间,据戎嘉余等(2012)的推断是在中—晚埃隆期(Middle to Late Aeronian),即位于石牛栏组与韩家店组之间。根据上述情况推断,宜昌地区在沉积龙马溪组黄绿色页岩段下部至上部的下部期间(相当于 *arcuata* 带下部至中部)所发生的兰多维列世第一幕大海退地质事件与黔北一带的"桐梓上升"地质事件应为同一地史时期的地质事件,这应引起高度的重视。

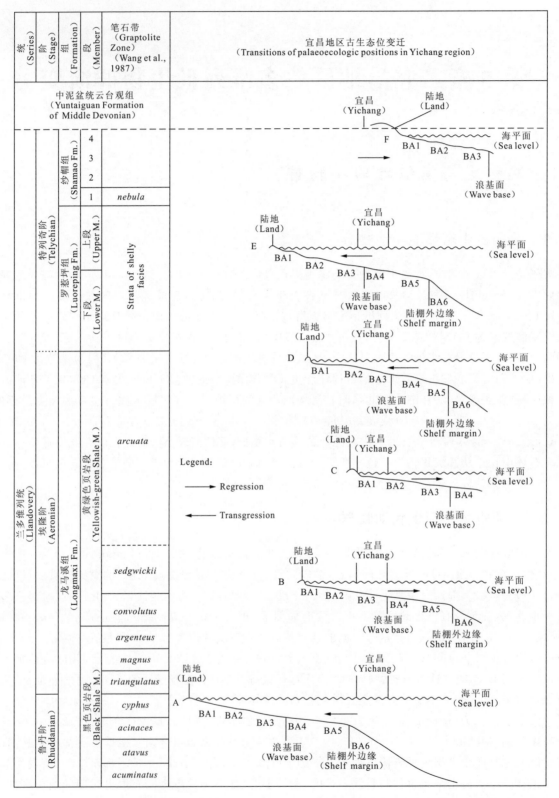

插图 5　宜昌地区兰多维列世古生态位变迁示意图

Illustr. 5　A sketch map showing transitions of the palaeoecologic positions during Llandoverian in Yichang region

5.2 沉积罗惹坪组时生物群大复苏

5.2.1 沉积罗惹坪组时的古生态环境

由于宜昌地区兰多维列世(Llandoverian)发生第二幕大海侵,其所引来的为暖水型海水,这与宜昌地区兰多维列世发生的第一幕大海侵时所引来的为冷水型的海水截然不同。因此其古生态环境也发生了质的变化,营造了极为适宜于底栖动物生存与繁衍的一种生态环境,从而产生了新的生物群。由于当时海侵不断地加大,到沉积罗惹坪组下段底部时,其生态底域位置从其前沉积龙马溪组黄绿色页岩段上部的BA1("*Lingula*"群落)→BA2(*Nucleospira*群落)的生态位置演变成BA3。而此时的海水所含$CaCO_3$可以部分沉淀,底栖动物能够从水体中吸收到足够的钙质去营造它们的介壳或骨骼。因此罗惹坪组下段最底部的岩层出现了薄层腕足类泥灰岩,在地貌上经常形成30～50cm高的小陡坎。其中的腕足类为*Sinokulumbella*(原来的*Stricklandia*或*Kulumbella*),是宜昌地区出现在最低的层位,并且密集成层,但其厚度仅有2～3cm。在*Sinokulumbella*岩层之上,紧接着出现极为丰富、贝体很小的*Meifodia*和少量的*Spinochonetes*等腕足类,并与少量小型单体珊瑚*Crassilasma*以及少量的三叶虫*Latiproetus*等化石共存。此介壳层是罗惹坪组下段底界的标志层,对地层划分与对比,以及对古生态环境的分析都具有重要意义。但由于该介壳层仅有10cm厚,在野外往往被忽视。前述的*Sinokulumbella*和*Spinochonetes*两个属本来应分别是BA4和BA4—BA5生态底域位置的典型代表。但考虑到它们是覆在*Nucleospira*群落(BA2)之上不太远的层位中出现,推断它们也可以在BA3生态位中少量生存。因此该介壳层所处的生态位被确定为BA3,并选用*Meifodia*作为典型代表(曾庆銮等,2015b)。另外,这一海侵与埃隆晚期全球海平面处于上升期应该有着密切关系,应引起高度的重视。

再往上到沉积罗惹坪组下段中部时,当时水介质中钙质含量越来越多,各门类化石越来越丰富,尤其是腕足类、珊瑚类、三叶虫类等门类化石特别丰富,在许多地层段内经常分别密集成层。其中的腕足类主要是以*Spinochonetes*-*Spinolella*-"*Clorinda*"群落为代表,前两者是生态位BA4—BA5的象征(曾庆銮等,2015b),而后者"*Clorinda*"的化石不多,而且还可能为*Brevilamnulella*,但它是BA5的典型代表(Ziegler,1965;Boucot,1975;戎嘉余,1986;戎嘉余等,1984)。根据本区的具体情况,将它们的生态位置确定为BA4—BA5。

再往上到沉积罗惹坪组下段上部至罗惹坪组上段时的腕足类群落一直是由*Sulcipentamerus*群落与*Sinokulumbella*群落频繁相互更替出现,其次数可达9次之多,时而出现*Sinokulumbella*灰岩,时而又出现*Sulcipentamerus*灰岩,但越往上*Sinokulumbella*灰岩越少,甚至完全被*Sulcipentamerus*灰岩取代。*Sulcipentamerus*群落是BA3生态位的典型代表,而*Sinokulumbella*群落则是BA4生态位的典型代表(Ziegler,1965;Boucot,1975;戎嘉余,1986)。从上表明进入罗惹坪组下段上部至上段沉积时期,宜昌地区当时所处的海域总体来说是相对较为稳定的,仅到罗惹坪组上段顶部时,其生态位才完全演变为BA3。

另外值得一提的是,宜昌地区进入沉积罗惹坪组时,单体珊瑚和小型复体珊瑚礁极为昌盛,表明当时的生态环境是非常适宜于珊瑚动物的繁衍与生长,并且经常与*Sulcipentamerus*群落共生。而珊瑚是海生底栖固着动物,喜欢生存在海水清澈、含盐度正常(35‰)、暖和海水(23～25℃)、海水深度约在50～70m之间的一种海底域。这不仅证明当时的海水是属于暖水型的,而且也可以印证*Sulcipentamerus*群落的生态位被确认为BA3是正确的。

5.2.2 沉积罗惹坪组时生物群大复苏

沉积罗惹坪组时,正好是宜昌地区在晚兰多维列世早期(Late Llandoverian early)发生兰多维列世

第二幕大海侵，其所引来的海水是暖水型的，其中的钙质含量又很适度，为海生底栖动物的繁衍与生长营造了极为优良的生态环境，从而导致当时生物群面貌焕然一新，各门类底栖动物呈现出欣欣向荣的景象。这堪比当今"大地回春，万物复苏"的情景。据目前的资料所知，当时门类多样性激增到 10 个，这与其前期仅有浮游笔石一个门类大为不同。当时生物群的最大特征是钙质壳的腕足动物重新兴旺到占主导地位，以及在宜昌地区首次出现暖水型、钙质骨骼的珊瑚类动物群也占有极其重要的地位；但前期占主导地位的几丁质骨骼的浮游笔石动物群却极为衰落，甚至见不到踪影。据曾庆銮等(2015b)对本区罗惹坪组主要 6 个门类属级兴衰的定量统计，各门类具体情况如下：

(1) 腕足类的新生属(含外来属)有 27 个，属的总数为 28 个(含延续属)，并以 $Meifodia$, $Sinokulumbella$, $Sinostricklandiella$, $Sulcipentamerus$, $Spinochonetes$, $Spinolella$, $Merciella$, $Aegiromena$, $Lissatrypa$, $Eospirifer$, $Zygospiraella$, $Salopina$, $Isorthis$, $Pleurodium$ 等重要属为代表。

(2) 珊瑚类的新生属(含外来属)有 19 个，属总数也是 19 个，并以 $Classilasma$, $Pycnactis$, $Pterophrentis$, $Streptelasma$, $Palaeophyllum$, $Protocystiphyllum$, $Cysticonophyllum$, $Tryplasma$, $Teratophyllum$, $Densiphylloides$, $Onychophyllum$, $Palaeofavosites$, $Mesofavosites$, $Favosites$, $Heliolitella$ 等属为代表。

(3) 牙形石的新生属(含外来属)23 个，属总数也是 23 个，并以 $Panderodus\ unicostatus$ (Branson et Mehl), $Walliserodus\ curvatus$ (Branson et Mehl), $Distomodus\ kentuckyensis$ (Branson et Mehl), $Synprioniodina$ cf. $silurica$ Walliser, $Dapsilodus\ obliquicostatus$ (Branson et Mehl)等分子为代表。

(4) 三叶虫类的新生属(含外来属)9 个，属总数(含延续属)11 个，并以 $Ptilillaenus$, $Shiqiania$, $Encrinuroides$, $Scotoharpes$, $Kosovopeltis$, $Latiproetus$, $Scharyia$, $Gaotania$, $Songkania$ 等属为代表。

(5) 笔石类的新生属(含外来属)2 个，属总数 11 个(含延续属)，并以 $Monoclimacis\ arcuata$, $Pristiograptus\ regularis$, $Diplagraptus\ atopus$, $Pseudoglyptograptus\ retroversus$ 等分子为代表，而且都是零星分布在罗惹坪组下段下部至中部。

(6) 头足类的新生属(含外来属)4 个，属总数也是 4 个，并以 $Yichangoceras$, $Eridites$, $Harrisoceras$, $Mixosiphonocerina$ 等为代表，而且只零星分布在罗惹坪组下段中部。

除上述主要 6 个门类外，几丁虫这一门类也已引起了人们高度的重视，但目前的材料不多，仅从耿良玉等(1988)的研究资料得知，在相当于罗惹坪组下段产有以 $Conochitina\ rossica$, $Conochitina\ iklaensis$, $Conochitina\ emmastensis$ 等分子为代表的几丁虫群。目前陈孝红、张淼等正在对其进行更深入的研究。另外在罗惹坪组下段还经常见有 $Pentagonocyclicus$(中央孔为五角形)和 $Cyclocyclicus$(中央孔为圆形)2 个属为代表的海百合这一个门类，可惜目前还很少有人去研究。

5.3 上扬子海走向消亡

5.3.1 沉积物和接触关系

沉积纱帽组时的沉积物或腕足类群落演替的情况都反映出宜昌地区开始发生兰多维列世第二幕特大级别的大海退。当时的沉积物和腕足类群落从下到上大致如下：

纱帽组第一段的岩性为黄绿色粉砂质泥岩夹薄层泥质粉砂岩，越往上粉砂质越多，与其下伏的"上五房贝石灰岩"层(属于 BA3)截然不同。当中产有腕足类 $Isorthis$, $Katastrophomena$, $Salopina$, $Nucleospira$, $Aegiromena$ 和少量笔石 $Climacograptus\ nebula$, $Oktavites\ planus$, $Pristiograptus\ varibilis$ 等分子。其中腕足类 $Nucleospira$ 的生态位为 BA2(戎嘉余等，1984)，而 $Katastrophomena$ 的生态位也为 BA2(曾庆銮等，2015b)。因此纱帽组第一段的生态位置应为 BA2。本段厚 77.4m。

纱帽组第二段,其岩性为黄绿色细砂岩与粉砂岩互层,岩性硬度大,在地貌上形成大陡坎。层面上经常出现波痕、雨痕、泥裂。本段产有少量腕足类 *Nucleospira*, *Isorthis*, *Striispirifer*, 以及少量三叶虫 *Latiproetus*, *Encrinuroides* 和稀少的笔石 *Pristiograptus regelaris*, *P. varibilis*, *Monograptus marri* 等类型化石。其中的腕足类 *Nucleospira* 是生态底域位置 BA2 的典型代表(戎嘉余等,1984;曾庆銮等,2015b)。但考虑到本段岩层表面有雨痕、泥裂等构造,因此沉积本段时的生态位为 BA1—BA2 较为合适,而且还经常露出水面,表明当时的水体很浅,上下波动较大。本段厚 125.8m。

纱帽组第三段,其岩性为灰绿色粉砂质泥岩夹细砂岩。当中产稀少腕足类 *Nucleospira*, *Nalivkinia* 和个别三叶虫 *Coronocephalus* 等化石,但都保存很差。其中 *Nalivkinia* 是生态位 BA1 的典型代表(戎嘉余等,1984),表明沉积本段时的水体很浅。本段厚 282m。

纱帽组第四段,其岩性为灰绿色中厚层细粒石英砂岩夹薄层粉砂岩。本段岩性硬度大,在地貌上形成大陡坎,当中大型斜层理或交错斜层理很发育,无化石,厚 185.3m,并与上覆中泥盆统云台观组灰白色厚层含砾石英砂岩呈平行不整合接触。

5.3.2 上扬子海走向消亡的证据

从前面纱帽组第 1—4 段的岩性来看,当时的海域是接受一套快速沉积,总厚度达 670.5m 的细碎屑岩,而且是属于由细到粗倒粒序的沉积过程。在第二岩性段的层面上还经常出现雨点痕和泥裂,表明当时有经常露出水面的情况。到第四段的细砂岩层,其大型斜层理和交错斜层理很发育,无化石,表明当时宜昌地区已处在滨海区,而且有可能是属于大河口处的滨海区,并从其周围快速上升的剥蚀区运来大量细碎屑物质在本区快速沉积。

从沉积纱帽组时的沉积物质和层面构造所反映的大海退还可以从当时生存的腕足类群落演替情况得到有力的印证。在罗惹坪组顶部生存的 *Sulcipentamerus* 群落,其生态位为 BA3。往上进入沉积纱帽组第一段时,生存的腕足类更替为 *Katastrophomena* 群落,其生态位为 BA2。再往上进入沉积纱帽组第二段时,生存的腕足类演变为 *Nucleospira* 群落,其生态位为 BA1—BA2。到沉积纱帽组第三段时,生存的腕足类又演替为 *Nalivkinia* 群落,其生态位为 BA1。最后到沉积纱帽组第四段时,已无化石,此时连生存于 BA1 生态位的 *Nalivkinia* 群落的腕足类也无法生存,从中暗示当时的海水可能快要退尽。再到纱帽组第四段顶界则与其上覆中泥盆统云台观组白色含砾石英砂岩呈平行不整合接触。由此表明到沉积纱帽组第四段顶界之后,宜昌地区已上升为陆,并成为剥蚀区。而这一特大级别的大海退地质事件,其力度大、范围广(中南地区区域地层表编写小组,1974),并被称为扬子上升(Yangtze uplift)(戎嘉余等,2012),从而几乎终止了我国南方自震旦纪陡山沱期早期海侵以来所形成的,并跨越整个震旦纪、寒武纪、奥陶纪,最后进入志留纪晚兰多维列世早期,历经时间长达约 2.72 亿年的上扬子海的生命。

说明:宜昌地区兰多维列统含有非常丰富的各门类化石,但考虑到已有 90 年的研究历史,各类论文和专著很多,目前仅对腕足类化石的一些新材料粘贴一些图版并进行描述。

第6章 化石系统描述

（所有标本都保存在中国地质调查局武汉地质调查中心）

6.1 赫南特动物群的化石

腕足动物门 Phylum Brachiopoda Duméril, 1806

  舌形贝亚门 Subphylm Linguliformea Williams et others, 1996

    舌形贝纲 Class Lingulata Gorjansky et Popov, 1985

      舌形贝目 Order Lingulida Waagen, 1885

        平圆贝超科 Superfamily Discinoidea Gray, 1840

          平圆贝科 Family Discinoidae Gray, 1840

圆凸贝属 Genus *Orbiculoidea* Orbigny, 1847

1847 *Orbiculoidea* Orbigny. P. 269.
1965 *Orbiculoidea* Orbigny; Moore. H285.
1966 *Orbiculoidea* Orbigny; 王钰等。102 页。
2000 *Orbiculoidea* Orbigny; Holmer et Popov. P. 90.

属型种：Genotype *Orbicula forbesii* Davidson, 1848.

特征简述：贝体轮廓亚圆形，同心生长纹发育；贝体侧视强烈背双凸型到凸平型；背壳呈圆锥状至亚圆锥状；腹壳呈低亚圆锥状至缓凹，后坡具长短不一的窄茎沟。

分布及时代：世界各地；奥陶纪至三叠纪。

圆凸贝？未定种 *Orbiculoidea*? sp.

图版（pl.）1，图（figs.）1-3

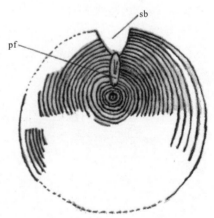

插图 6 *Orbiculoidea*? sp. 的腹内模
（据图版1，图1）
Illustr. 6 Ventral internal mold of *Orbiculoidea*? sp. (from pl. 1, fig. 1)
pf-茎沟（pedical furrow）；sb-裂口（schizo breach）

描述：当前仅获得2枚不完整腹内模和一枚存疑背外（HB710的壳长约9mm，壳宽8.2mm；HB699的壳长约9mm；HB351的壳长11mm，壳宽9mm）。贝体小，腹壳轮廓呈亚圆形，凸度低缓；腹喙小，位于腹壳中后部；同心纹发育，围绕着喙部呈同心圆状生长，开始的同心纹圆而细密，随着贝体的增大，同心纹的间距也随着增宽；腹壳后坡具一窄茎沟；在茎沟的前端开始张裂，形成约50°角的裂口；同心纹都不穿越茎沟和裂口（图版1，图1；插图6）。背壳？凸度低；背喙位于背壳顶偏后，轻微向后弯曲；背壳同心纹也很发育。

讨论：从当前贝体的形态、同心生长纹很发育以及具显著的茎沟来看，似乎应归于 *Orbiculoidea* Orbigny（1847）；但从腹壳茎沟的前端开始、腹壳后坡张裂成约50°角的裂口来看，它又类似于

Discinisca Dall(1871b)。由于目前的标本太少，保存又欠佳，内部构造也不清楚，因此暂时疑问归于 *Orbiculoidea* Orbigny，不排除是属于新的一种类型。

产地层位：湖北宜昌王家湾、丁家坡；奥陶系顶部五峰组观音桥段(赫南特阶中部)。

骷髅贝亚门 Subphylum Craniiformea Popov et others,1993
　骷髅贝纲 Class Craniata Williams et others,1996
　　非骷髅贝目 Order Craniopsida Gorjansky et Popov,1985
　　　非骷髅贝超科 Superfamily Craniopsoidea Williams,1963
　　　　非骷髅贝科 Family Craniopsidae Williams,1963

非骷髅贝属 Genus *Craniops* Hall,1859b

1859b　*Craniops* Hall. P. 84.
1965　*Craniops* Hall;Moore. H273.
2000　*Craniops* Hall;Popov et Holmer. P. 164.

属型种：Genotype ? *Orbicula squamiformis* Hall,1843.

特征简述：贝体轮廓长卵形，侧视不等双凸型；同心片层发育；两壳顶区肿胀；背、腹壳的内脏区都隆起形成低的平台；两壳内边缘都具有饰边。本属与 *Paracraniops* Williams(1963)的区别是后者的背内缺乏内脏平台。

分布及时代：世界各地；奥陶纪至泥盆纪。

可分非骷髅贝 *Craniops partibilis* (Rong)
图版(pl.)1,图(figs.)4-7

1979　"*Paracraniops*"*partibilis* Rong. 2页,图版1,图1,2。
1981　*Sanxiaella partibilis* (Rong);常美丽。558页,图版1,图1-6。
1983　*Sanxiaella partibilis* (Rong);曾庆銮。113页,图版13,图23-25。
2006　*Paracraniops* sp. 戎嘉余。297页。

描述：贝体极小，通常壳长在 2mm 以内，壳宽很少超过 1.5mm；轮廓长卵形，侧视低缓双凸型，壳表饰显著同心纹。腹壳长度稍微大于背壳长度，凸度低缓，但稍微大于背壳；前缘和两侧缘都呈宽弧状，但两后侧缘迅速向腹喙部尖缩；腹假铰合面中等长。背壳凸度更低缓，背假铰合面更短。背、腹壳内部边缘都具显著的围边(marginal limbus)。背、腹壳内部的内脏区都有显著隆起的内脏台(visceral platform)(图版1,图6,7;插图7)(这些特征，在常美丽,1981年,图版1,图1-4呈现的更为清晰)。

讨论：戎嘉余(1979)将本种带引号归入"*Paracraniops*"这个属；戎嘉余和常美丽(1981)以 "*Paracraniops*" *partibilis* Rong (1979)为模式种建立 *Sanxiaella* 一属(常美丽,1981)。从目前在同一地区同一个层位中

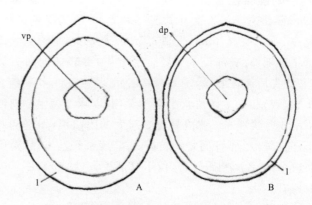

插图 7　*Craniops partibilis*(Rong)的内部构造
A-腹内模(据图版1,图6);B-背内模(据图版1,图7)
Illustr. 7　Interior structures of *Craniops partibilis*(Rong)
A-Ventral internal mold(from pl. 1,fig. 6);B-Dorsal internal mold(from pl. 1,fig. 7)
dp-背内脏台(dorsal visceral platform);l-围边(marginal limbus);vp-腹内脏台(ventral visceral platform)

获得该种大量的背、腹内模及外模来看,该种的腹内具有明显的腹内脏台(ventral visceral platform),在其背内也具有显著的背内脏台(dorsal visceral platform),这与常美丽(1981),图版1,图1-4完全一致。另外,在其背、腹内部边缘都具显著的围边(marginal limbus)。上述这些重要特征与 Craniops Hall (1859b)的属征相符合,而与 Paracraniops Williams(1963)的背内缺乏背内脏台明显不同。因此"Paracraniops" partibilis Rong(1979)应归入 Craniops Hall (1859b)这个属,而 Sanxiaella Rong et Chang (1981)应是 Craniops Hall(1859b)的同义名。

产地层位:湖北宜昌王家湾、丁家坡、黄花场;奥陶系顶部五峰组观音桥段(赫南特阶中部)。

骷髅贝目 Order Craniida Waagen,1885

  骷髅贝超科 Superfamily Cranioidea Menke,1828

  骷髅贝科 Family Craniidae Menke,1828

刺骷髅贝属 Genus *Acanthocrania* Williams,1943

1943 *Acanthocrania* Williams. P. 71.
1965 *Acanthocrania* Williams;Moore. H290.
2000 *Acanthocrania* Williams;Bassett. P. 171.

属型种:Genotype *Crania spiculata* Rowley,1908.

特征简述:背壳呈锥状至亚锥状;壳表具许多细乳头状凸起和许多倾覆状中空小刺,并以喙部为中心向周围呈放射状展布;背喙部位于背壳中后部;背壳后坡较陡峻;前对闭肌痕位于背喙部前坡上方,通常大于后对闭肌痕;而后对闭肌痕则位于背喙部后坡。腹壳情况不明。

分布及时代:世界各地;奥陶纪至石炭纪。

宜昌刺骷髅贝 *Acanthocrania yichangensis* Zeng

图版(pl.)1,图(figs.)8a-8d

1983 *Acanthocrania yichangensis* Zeng. 114页,图版17,图30-33。

描述:目前仅获得一枚很好的背内模(曾庆銮于1983年将它误认为腹内模,应予修正)。背壳呈低锥状,壳宽7mm,壳长6mm,壳高2.5mm;轮廓亚圆形,最大高度位于喙部;喙部位于后铰合缘上方,并轻微向后方倾斜;背壳后坡较陡峻,前坡缓斜。壳表饰有许多微小乳头状凸起和许多倾覆状中空小刺,并以喙部为中心向四周呈放射状展布。背内具有2对大的闭肌痕,均呈椭圆形;前对较大,彼此间相距较近,位于喙部稍前方;后对闭肌较小,两者间相距较远,位于喙部后方(图版1,图8a-8d;插图8)。

比较:*Acanthocrania yichangensis* Zeng 与 *A. setigera*(Hall)(Cooper,1956,P. 285,pl. 25,figs. 1-5)较相似,但前者的背壳喙部位于后铰合缘上方,并轻微向后方倾斜,后坡较陡峻,而前坡较长,缓斜,前对闭肌痕较大等特征,明显可和 *A. setigera*(Hall)相区分。

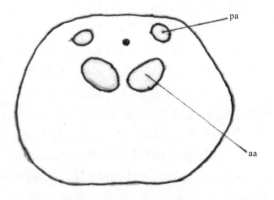

插图8 *Acanthocrania yichangensis* Zeng 的背内模
(据图版1,图8a)

Illustr. 8 Dorsal internal mold of *Acanthocrania yichangensis* Zeng(from pl. 1,fig. 8a)

aa-前闭肌痕(anterior adductor scars);pa-后闭肌痕(posterior adductor scars)

产地层位:湖北宜昌黄花场;上奥陶统顶部五峰组观音桥段(赫南特阶中部)。

友基贝属 Genus *Philhedra* Koken,1889

1889　*Philhedra* Koken. P. 465.
1965　*Philhedra* Koken;Moore. H291.
1968　*Philhedra* Koken;Bergström. P. 7.
1987　*Philhedra* Koken;Temple. P. 28.
2000　*Philhedra* Koken;Bassett. P. 182.

属型种:Genotype *Philhedra baltica* Koken,1889.

特征简述:背壳呈亚圆锥形;喙部小,位于壳顶中后部;后坡陡峻,后边缘微凹,但有的呈圆状;壳表饰粗大、形成直行并且呈放射状排列的中空壳刺,但有的种近呈放射纹状;具同心层或同心生长纹。背肌痕面和腹壳的情况至今不了解。

分布及时代:亚洲、北美和欧洲;奥陶纪。

友基贝未定种 *Philhedra* sp.
图版(pl.)1,图(figs.)9-11

描述:贝体很小,背壳高度通常 2~3.2mm,宽 3.2~3.6mm;轮廓呈亚圆锥状;喙部小,位于壳顶后部;背壳最大凸度位于喙部,并从喙部逐步向四周降低呈斜坡状。壳表饰呈不很规则的放射纹状,在放射纹状上隐约见到小刺痕;同心纹不发育。

讨论:目前采到多枚背壳和背外模(曾庆銮于 1983 年误认为腹内模,应予以修正,115 页,图版 13,图 10,11;图版 17,图 28,29)。当前标本与 Cooper(1965),P. 291,pl. 26,figs. 4,5 的 *Philhedra depressa* Cooper 和 Bergström(1968) 的 *Philhedra* sp. (pl. 1, fig. 6) 以及 Temple(1987),pl. 1,fig. 10 的 *Philhedra* cf. *grayii*(Davidsom)都很接近,但标本都不太好,暂时定为 *Philhedra* sp. 较为合适。但应该指出:常美丽(1981)采于宜昌王家湾五峰组观音桥段的 *Philhedrella wangjiaensis* Chang,图版 1,图 8 的标本与当前的标本没有什么区别,而且仅有一枚不太清楚的背内模(当时常美丽误认为腹内模),建议改归于 *Philhedra* sp. 较为合适。

产地层位:湖北宜昌王家湾、丁家坡;上奥陶统顶部五峰组观音桥段(赫南特阶中部)。

小嘴贝亚门 Subphylum Rhynchonelliformea Williams et others,1996
　扭月贝纲 Class Strophomenata Williams et others,1996
　　扭月贝目 Order Strophomenida Öpik,1934
　　　扭月贝超科 Superfamily Strophomenoidea King,1846
　　　　扭月贝科 Family Strophomenidae King,1846
　　　　　叉形贝亚科 Subfamily Furcitellinae Williams,1965

小月贝属(新属)Genus *Minutomena* Zeng,Zhang et Han(gen. nov.)

属型种:Genotype *Minutomena yichangensis* Zeng,Zhang et Han(gen. et sp. nov.).

词源:Minuto(拉丁文),微小的,表示新属为小型的扭月贝类。

特征简述:壳小,轮廓亚方圆形,侧视低缓双凸型或轻微颠倒型;背、腹三角孔覆隆起的背三角板和腹假三角板。壳表饰不等放射线和少量微弱同心纹。腹内齿板强壮,内弯,延伸至腹肌痕面两外侧。背内主突起短粗,双叶型;铰窝大,近三角状;内铰窝脊强烈异向张开;背壳底两中后侧各具一根侧脊,但缺失背中隔板。

描述:壳小,壳宽通常在 7mm 左右;轮廓亚方圆形;侧视低缓双凸型或轻微颠倒型;铰合线直,稍微短于最大壳宽;主端钝角状或近直角状。腹壳凸度低缓,仅在喙部附近稍微隆起,有时在腹壳中前部轻微凹下；腹喙中等大，不弯曲；腹铰合面中等发育，强烈斜倾型；腹三角孔被拱起的假三角板覆盖。背壳凸度低缓，仅在近前缘处轻微隆起，而在喙部前方低平或轻微凹下；背喙小；背三角孔被背三角板覆盖。壳表饰不等放射线，一般作二次分叉，线顶圆滑；同心纹少，而且微弱，偶尔在背壳中前部具少许同心层。假疹壳。

腹内：铰齿粗壮，三角脊状；齿板发育，内弯，延伸在横宽腹肌痕两外侧（图版2，图2a，2b）；腹肌痕面模糊不清。

背内：主突起短粗，双叶型；铰窝大，近三角状；内铰窝脊低宽，强烈异向展伸（图版2，图3）；在背壳底两中后侧各具一根粗壮侧脊（图版2，图3，4a，2b；插图9）。

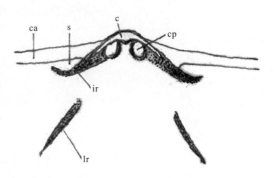

插图9 *Minutomena*（gen. nov.）的背壳内部构造（据图版2，图4a，2b）

Illustr. 9 Interior structures of dorsal valve of *Minutomena*(gen. nov.) (from pl. 2, figs. 4a, 2b)

c-背三角板(chilidial plates); ca-铰合面(cardinal area); cp-主突起(cardinal process); ir-内铰窝脊(inner socket ridge); lr-侧脊(lateral ridge); s-铰窝(socket)

比较：*Minutomena*(gen. nov.)的外形、腹内构造，以及背内的主突起形状和2根侧脊的情况都与 *Furcitella* Cooper(1956)很相似，但是 *Furcitella* 的背内具有分叉状的背中隔脊(Cooper, 1956, pl. 229, figs. 20, 23, 24, 26)，而 *Minutomena*(gen. nov.)的背内缺失背中隔板，两者易于区分。另外，因本新属背内具一对显著侧脊，它与 Subfamily Strophomeinae 亚科的含义不同，而却与 Subfamily Furcitellinae 亚科的含义相符合，因此将本新属归入 Subfamily Furcitellinae Williams(1965)这个亚科内。

分布及时代：中国中南部；晚奥陶世赫南特期(Hirnantian)至志留纪兰多维列世(Llandoverian)。

宜昌小月贝属（新属、新种）*Minutomena yichangensis* Zeng, Zhang et Han (gen. et sp. nov.)

图版(pl.)2，图(figs.)1-5

词源：Yichang(汉语拼音)，为化石产地。

描述：贝体小（表1），轮廓亚圆形；侧视低缓双凸型或轻微颠倒型；铰合线直，稍微短于或近等于最大壳宽；主端钝角状或近直角状；侧缘和前缘都呈宽弧状。腹壳凸度低缓，最大凸度位于喙部附近，在壳表中前部有时轻微凹下；腹喙较大，但不高；腹铰合面适度高，强烈斜倾型；腹三角孔被拱形状假三角板覆盖。背壳凸度低平，有时在壳表中前部轻微隆起，但在背喙前方反而轻微凹下；背喙小；背铰合面极低。壳表饰低圆放射线，一般作2次分叉，线顶圆滑；具少许微弱同心纹或偶见同心层。假疹壳。

表1 宜昌小月贝介壳测量(单位:mm)

Table 1 Shell measurements of *Minutomena yichangensis*(gen. et sp. nov.)(in mm)

采集号 (Coll. No.)	登记号 (Cat. No.)	腹壳(ventral valve)		背壳(dorsal valve)		备注 (remarks)
		长(length)	宽(width)	长(length)	宽(width)	
WH2	HB677	8.2	10.5			副型(paratype)
Pm065-6-1F	YB2	4.9	约7			副型(paratype)
Pm065-6-1F	YB1			5	9	副型(paratype)
DH2	HB678			4.6	5.2	正型(holotype)

背、腹内部构造同属征。

产地层位：湖北宜昌丁家坡、杨家湾；上奥陶统顶部五峰组观音桥段至志留系兰多维列统罗惹坪组下段中部。

瑞芬贝科 Family Rafinesquinidae Schuchert,1893

薄皱贝亚科 Subfamily Leptaeninae Hall et Clarke,1894

薄皱贝属 Genus *Leptaena* Dalman,1828

1828 *Leptaena* Dalman. P. 93.
1956 *Leptaena* Dalman；Cooper. P. 820.
1957 *Leptaena* Dalman. Spjeldnaes. P. 171.
1962 *Leptaena* Dalman；Williams. P. 197.
1965 *Leptaena* Dalman；Willams. H391.
1966 *Leptaena* Dalman；王钰、金玉玕、方大卫。300 页。
1967 *Leptaena* Dalman；Havlíček. P. 86.
1968 *Leptaena* Dalman；Bergström. P. 14.
1968 *Leptaena* Dalman；Cocks. P. 299.
1973 *Leptaena* Dalman；Boucot. P. 20.
1974 *Leptaena* Dalman；Williams. P. 148.
1977 *Leptaena* Dalman；Bassett. P. 123.
1980 *Leptaena* Dalman；Nikitin. P. 56.
1983 *Leptaena* Dalman；曾庆銮。120 页。
1984 *Leptaenopma* Marek et Havlíček；Rong. P. 154.
1997 *Leptaena* Dalman；曾庆銮。9 页。
2000 *Leptaena* Dalman；Cocks et Rong. P. 243.

属型种：Genotype *Leptaena rugosa* Dalman,1828.

特征简述：贝体中等大，体腔区窄；轮廓亚方形，铰合线直，等于最大壳宽；侧视凹凸型。腹壳顶区平坦，在两侧边和近前缘处都向背方膝曲呈陡坡，但拖曳部不长。腹三角孔被假三角板覆盖。背壳浅凹，中部宽平，两侧边和近前缘处均向上（背方）折曲，折边陡峻，但不太高；背三角孔被背三角板覆盖。壳表饰细密放射纹，同心皱发育。

铰齿粗壮；齿板发育，内弯，延伸在腹肌痕面两外侧；腹肌痕面大，呈亚圆形至长卵形。

背窗台窄小；铰窝三角状；主突起密集双叶型；背肌痕面大，明显四分；前对闭肌痕较小，位于背中隔板后方；后对闭肌痕较大，位于前对闭肌痕两后侧；在背肌痕面中前方具一背中隔板（插图 10）。

分布及时代：世界各地；中奥陶世至志留纪。

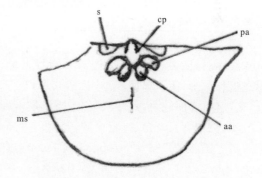

插图 10 *Leptaena* Dalman 的背内构造
（据 2000，Treatise，P. 244，fig. 150，1d）
Illustr. 10 Interior structures of dorsal valve of *Leptaena* Dalman(from 2000,Treatise,P. 244,fig. 150,1d)
aa -前闭肌痕（anterior adductor scars）；cp -双叶型主突起（bilobed cardinal process）；ms -中隔板（median septum）；pa -后闭肌痕（posterior adductor scars）；s -铰窝（socket）

黄花薄皱贝 *Leptaena huanghuaensis* Zeng

图版(pl.)2,图(figs.)6-10;图版(pl.)3,图(figs.)1-8;
图版(pl.)49,图(fig.)1

1983 *Leptaena huanghuaensis* Zeng.120页,图版16,图1-6。
1984 *Leptaenopoma trifidum* Marek et Havlíček;Rong.P.154,pl.11,figs.16,18-21.
2006 *Leptaena trifidum* (Marek et Havlíček);戎.294页,图版2,图19。

补充描述:当前在和以前描述标本的同一地区、同一层位中采到更多更好的标本,有必要补充描述:贝体较大,通常壳宽 24~35mm,壳长 16.5~22mm(表2);轮廓近横方形;铰合线直,近等于最大壳宽;主端经常呈小耳状,耳翼稍微尖突(图版3,图1a,4);侧视轻微凹凸型,体腔区贝体薄。腹喙大,不弯曲;腹三角孔被拱形状假三角板覆盖;腹铰合面中等高,斜倾型,饰有横纹;腹壳顶区平坦,两侧区和近前缘处都向背方膝曲,侧坡和前坡都形成较陡的斜坡状(图版3,图4)。背壳浅凹,壳表中部平坦,在两侧区和近前缘处都向上折曲,折边较陡峻,使整个背壳犹如半个浅盘状;背喙不显著;背三角孔被背三角板覆盖。壳表饰细密放射纹;同心皱发育。假疹壳。

表2 黄花薄皱贝介壳测量(单位:mm)
Table 2 Shell measurements of *Leptaena huanghuaensis* Zeng (in mm)

采集号 (Coll. No.)	登记号 (Cat. No.)	腹壳(ventral valve)		背壳(dorsal valve)	
		长(length)	宽(width)	长(length)	宽(width)
WH2	HB69	约9	23		
DH2	HB322	13	24		
WH2	HB283	22	35		
DH2	HB357	32	37		
WH3	HB706	19	27		
HH2	IV45843			16.5	24
DH3	HB305a			17	28

腹内:铰齿粗壮(图版3,图1c);齿板发育,内弯,延伸在腹肌痕面两外侧;腹肌痕面大而显著;亚圆形或近长卵形;闭肌痕面较小,长椭圆形,居于中间,中线被显著或不显著中肌隔分成两半(图版3,图1a,3);启肌痕面较大,月牙状,位于闭肌痕两外侧,并包围着闭肌痕(图版3,图1a)。

背内:铰窝长三角状;背窗腔较窄小;主突起短粗、紧密双叶型(图版3,图6),有时在主突起后方具有一对方圆状的副铰窝(secondary sockets)(图版3,图8c;图版49,图1)(注:此名来自Moore,1965,H364,fig.232,1d);背肌痕面大,不易区分;在背肌痕中间具一显著短而粗的中肌隔,在中肌隔之前具一细弱背中隔板(图版3,图6);有时在背窗台之前的背壳底轻微隆起,形成三角状的背台,台的前缘平直,中间及其两前侧都被一低宽的短纵脊所支持(图版3,图5);在背台前方的背壳底也具有一微弱背中隔板。

讨论:从本种的主突起为短粗、紧密双叶型,以及背肌痕面和背中隔板性质来看,当前标本属于 *Leptaena* Dalman(1928)应是无疑的。但有时在主突起前方的背壳底轻微隆起形成雏形的三角形背台,背台前方具3根短脊,以及在主突起后方具有一对明显的副铰窝等特征又与 *Leptaena* 其他各个种有明显的差别。以上说明本种的特征是介于 *Leptaena* 和 *Laptaenopoma* Marek et Havlíček(1967)之间的类型,而且也表明 *Leptaenopma* 这个属应是从 *Leptaena* 演化而来的。

产地层位:湖北宜昌黄花场、丁家坡、王家湾;上奥陶统顶部五峰组观音桥段(赫南特阶中部)。

薄盖贝属 Genus *Leptaenopoma* Marek et Havliček,1967

1967 *Leptaenopoma* Marek et Havliček. P. 282.
1967 *Leptaenopoma* Marek et Havliček;Havliček. P. 88.
1968 *Leptaenopoma* Marek et Havliček;Bergström. P. 15.
1979 *Leptaenopoma* Marek et Havliček;戎嘉余。2页。
1980 *Leptaenopoma* Marek et Havliček;Nikitin. P. 58.
1981 *Leptaenopoma* Marek et Havliček;常美丽。562页。
1983 *Leptaenopoma* Marek et Havliček;曾庆銮。120页。
1984 *Leptaenopoma* Marek et Havliček;Rong. P. 153.
2000 *Leptaena* Dalman;Cocks et Rong. P. 243.

属型种:Genotype *Leptaenopoma trifidum* Marek et Havliček,1967.

修订后的特征简述:*Leptaenopoma* 的壳形像 *Leptaena* Dalman,其最大的差别是 *Leptaenopoma* 的背窗台宽大;主突起大,三叶型;背窗台之前的背壳底轻微隆起,形成极为强大、三角状背台,背台与背窗台相融合,台面很平坦;背台前缘平直;背台之前具一粗宽中隔脊;背台两前侧各有一根短侧隔脊;背中隔脊前方具一细弱背中隔板(图版 4,图 5a-5c;图版 5,图 7a,7b;图 49,图 2;插图 11-B);在老年壳的背壳底,于两侧隔脊之前还各具一微弱侧隔板,因此在老年壳底中部被分成四小块形状犹如两对背肌痕面(图版 4,图 6),其实不然,背肌痕面应在背台内,但不易区分(图版 49,图 2)。至于腹壳内部构造特征详见图版 4,图 2a,2b 和插图 11-A。

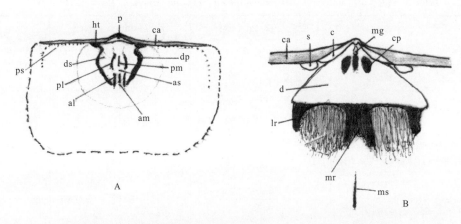

插图 11 *Leptaenopoma trifidum* Marek et Havliček 的内部构造
Illustr. 11 Interior structures of *Leptaenopoma trifidum* Marek et Havliček
A-腹内模(据图版 4,图 2a,2b);B-背内模(据图版 4,图 5c)
A - Ventral internal mold(from pl. 4,figs. 2a,2b);B-Dorsal internal mold(from pl. 4,fig. 5c)
al-前侧肌隔(anterior lateral myophragm);am-前中肌隔(anterior median myophragm);as-闭肌痕(adductor scars);c-背三角双板(chilidial plates);ca-铰合面(cardinal area);cp-主突起(cardinal process);d-背台(dorals platform);dp-齿板(dental plates);ds-启肌痕(diductor scars);ht-铰齿(hinge teeth);lr-侧脊(leteral ridges);mg-中沟(median groove);mr-中隔脊(median ridge);ms-中隔板(median septum);p-假三角板(pseudodeltidium);pl-后侧肌隔(posterior lateral myophragm);pm-后中肌隔(posterior median myophragm);ps-假疹壳细突起(tubercles of pseudopunctate shell);s-铰窝(socket)

讨论:上述 *Leptaenopoma* 那些重要特征在 Havliček(1967)pl. 13,figs. 4,6,8,9 同样有显露,只不过没有我们当前标本那样清晰而已。因此 *Leptaenopoma* 与 *Leptaena* 有着巨大的本质差别,绝对不可作为 *Leptaena* 的同异名,应恢复 *Leptaenopoma* Marek et Havliček(1967)这一属名。

宜昌地区观音桥段所产的 *Leptaenopoma* 具有 3 种完全不同的主基形态,可分为 *L. trifidum*,

L. yichangense 和 *L. rugosa* 三个种，其具体特征将分别简述于后。

分布及时代：中国中南部、波希米亚地区、瑞典中南部；晚奥陶世赫南特中期。

三分薄盖贝 *Leptaenopoma trifidum* Marek et Havliček

图版(pl.)4,图(figs.)1-9;图版(pl.)49,图(fig.)2

1967　*Leptaenopoma trifidum* Marek et Havliček. P. 283, pl. 4, figs. 1, 4, 8.
1967　*Leptaenopoma trifidum* Marek et Havliček; Havliček. P. 88, pl. 13, figs. 1-10, 16.
1977　*Leptaena rugosa* Dalman; Mitchell. P. 108, pl. 23, figs. 16, 19, 25.
1981　*Leptaenopoma trifidum* Marek et Havliček; Chang. 563 页, 图版 1, 图 32。
1983　*Leptaenopoma trifidum* ? Marek et Havliček; Zeng. 120 页, 图版 16, 图 16。
1984　*Leptaenopoma trifidum* Marek et Havliček; Rong. P. 154, pl. 11, figs. 11, 12, 14, 15.
2000　*Leptaena trifidum* (Marek et Havliček); Cocks et Rong. P. 243, fig. 150. 1k.

描述：贝体较大，通常壳长 14～18mm，壳宽 22～32mm（表 3）；轮廓近横方形；铰合线直，近等于最大壳宽；主端近直角状；贝体侧视凹凸型。体腔狭窄。腹壳凸度低，顶区宽阔、平坦，两侧边和近前缘处都向背方膝曲，拖曳部短，两侧坡和前坡形成较陡斜坡状；腹铰合面低，斜倾型；腹喙大，不弯曲；腹三角孔被隆起假三角板覆盖。背壳浅凹，壳表中部宽平，但两侧边和近前缘处都向上折曲，折边不高，但较陡；背铰合面低，正倾型；背三角孔被背三角板覆盖。壳表饰细圆放射纹和不太规则同心皱。假疹壳。

表 3　三分薄盖贝介壳测量（单位：mm）
Table 3　Shell measurements of *Leptaenopoma trifidum* Marek et Havliček (in mm)

采集号 (Coll. No.)	登记号 (Cat. No.)	腹壳(ventral valve)		背壳(dorsal valve)		备注 (remarks)
		长(length)	宽(width)	长(length)	宽(width)	
DH2	HB355	18.5	约 33			
DH2	HB309		约 19			
WH2	HB279	约 18	约 32			
DH2	HB292	17	24			
DH3	HB310				约 30	后选型(metatype)
DH3	HB491			约 18	约 25.8	
WH3	HB293				约 22	
WH2	Hb58			约 14		
DH3	HB312				约 14	

腹内：铰齿粗壮，呈低脊状；齿板发育，向前延伸在腹肌痕面两外侧，其前端不连接；腹肌痕面大，形态多变，从长卵形至亚圆形；一对闭肌痕面较小，居中，并以前、后两对侧肌隔与其外侧的启肌痕隔开；在闭肌痕纵中线前、后还具 2 根中肌隔将闭肌痕分成两半（图版 4，图 2b）；一对启肌痕较大，呈月牙状，位于闭肌痕两外侧，其前端与闭肌痕前端近等长。

背内：铰窝显著，呈椭圆状（图版 4，图 8）；背窗台宽平；主突起强大，三叶型；中主突起细柱状，相对较小，两侧主突起椭圆状，较粗壮；中突起之后具一显著中沟（图版 4，图 5c）；背窗台之前的背壳底轻微隆起形成宽阔、三角状的背台，背台前缘平直；背台之前具一低宽背中隔脊；背台两前侧各具一显著背侧隔脊（图版 4，图 5c；图版 49，图 2）；在背中隔脊之前不远还具一细中隔板；在老年壳底于细中隔板两外侧还各具一细侧隔板,连同前面的中隔脊和侧隔脊将背壳底中部分成前、后两对,其形态犹如两对背肌痕面(图版 4,图 6),其实不然;背肌痕面应位于背台内,但不易区分。

产地层位:湖北宜昌丁家坡、王家湾;上奥陶统顶部五峰组观音桥段(赫南特阶中部)。

皱纹薄盖贝 *Leptaenopoma rugosas* Zeng
图版(pl.)5,图(figs.)1-9;插图(Illustr.)12-B

1983 *Leptaenopoma rugosas* Zeng.121 页,图版 16,图 7-13。
1984 *Leptaenopoma trifidum* Marek et Havlíček;Rong.P.155.

描述:贝体较大,通常壳长 17~24mm,壳宽 25~36mm(表 4);轮廓近横方形;侧视凹凸型,体腔狭窄;铰合线直,近等于最大壳宽;主端尖翼状或近直角状。腹壳凸度低,中部壳面宽阔、平坦,但两侧边和近前缘处都向背方膝曲,拖曳部不长,侧坡和前坡都呈斜坡状;腹喙显著,不弯曲;腹铰合面中等高,斜倾型;腹三角孔被假三角板覆盖。背壳浅凹,壳表中部宽阔、平坦,但两侧边和近前缘处都向上折曲呈陡坡状,致使整个背壳犹如半个浅盘状;背铰合面低,正倾型;背三角孔被背三角板覆盖。壳表饰细圆放射纹和不规则同心皱。假疹壳。

表 4 皱纹薄盖贝介壳测量(单位:mm)
Table 4 Shell measurements of *Leptaenopoma rugosas* Zeng(in mm)

采集号 (Coll. No.)	登记号 (Cat. No.)	腹壳(ventral valve)		背壳(dorsal valve)		备注 (remarks)
		长(length)	宽(width)	长(length)	宽(width)	
WH2	HB287	19	30			
WH2	HB77	24	36			
WH3	HB288	约17	25			
DH2	HB308	23	33			
WH2	HB94				约24	
DH2	HB307					后选型(metatype)
WH2	HB323				约30	

腹内:铰齿低脊状;齿板发育,沿着腹肌痕面两外侧向前延伸,但其前端不会合;腹肌痕面大,长卵形;中肌隔较显著,不规则;侧肌隔微弱;启肌痕面较大,位于闭肌痕两外侧,但界线模糊;闭肌痕面较小,居于整个腹肌痕面中间。

背内:铰窝显著,亚三角状(图版 5,图 5);背窗台宽大;主突起非常发育,三叶型;中突起特别强大,卵圆形;两侧主突起相对较小;3 个主突起的前端都呈乳头状尖凸,并且排成一横排(图版 5,图 7a-7c);中突起之后无中沟;背窗台与其前方背壳底轻微隆起融合形成宽阔、三角状的背台,背台前缘直(图版 5,图 7b);背台前方被 3 根近平行的短隔脊所支持;背中隔脊之前不远还具一背中隔板(图版 5,图 6,8a,8b,9);在老年壳底于背中隔板两外侧还各具一侧隔板(图版 5,图 8a,8b)。背肌痕面位于背台内,但不易区分(图版 5,图 7b)。

比较:*Leptaenopoma rugosas* 与 *L.trifidum* 的区别是 *L.rugosas* 的中突起特别强大,呈卵圆形,而两侧主突起相对较小,并且 3 个主突起的前端形成一排乳头状的尖凸,另外,在中突起之后无中沟;而 *L.trifidum* 的中突起呈细柱状,远比两侧主突起细弱,另外,在中突起之后具有中沟(插图 12-A,B)。

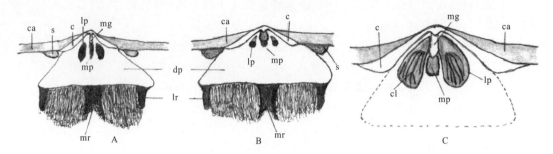

插图 12 背壳内部构造对比图

Illustr. 12 The interior structure comparisons of dorsal valves

A - *Leptaenopoma trifidum*(from pl. 4, fig. 5c); B - *Leptaenopoma rugosas*(from pl. 5, fig. 7b); C - *Leptaenopoma yichangense*(from pl. 6, fig. 5)

c -背三角板(chilidium); ca -铰合面(cardinal area); cl -主突起叶(cardinal process lobes); dp -背台(dorsal platform); lp -侧主突起(lateral cardinal process); lr -侧脊(lateral ridges); mg -中沟(median groove); mp -中主突起(median process); mr -中隔脊(median ridge); s -铰窝(socket)

产地层位:湖北宜昌王家湾、丁家坡;上奥陶统顶部五峰组观音桥段(赫南特阶中部)。

宜昌薄盖贝 *Leptaenopoma yichangense* Zeng

图版(pl.)6,图(figs.)1-9;插图(Illustr.)12-C

1983 *Leptaenopoma yichangense* Zeng. 120 页,图版 16,图 17-23。
1984 *Leptaenopoma trifidum*(Zeng); Rong. P. 155.

补充描述:贝体较大,一般壳长 19~23mm,壳宽 28~32mm(表 5);轮廓近横方形;贝体很薄,侧视凹凸型;铰合线直,近等于最大壳宽;主端尖翼状或近直角状。腹壳凸度低,顶区宽阔、平坦,两侧边和近前缘处都向背方膝曲,但拖曳部短;侧坡和前坡都呈斜坡状;腹喙大,不弯曲;腹铰合面中等高,斜倾型;腹三角孔被假三角板覆盖。背壳浅凹,壳表中部平坦,但两侧边和近前缘处都向上折起,折边不高,但陡峻;背铰合面低(图版 6,图 4b,4c),正倾型;背三角孔被背三角板覆盖。壳表饰细密放射纹和不规则同心皱。假疹壳。

表 5 宜昌薄盖贝介壳测量(单位:mm)

Table 5 Shell measurements of *Leptaenopoma yichangense* Zeng(in mm)

采集号 (Coll. No.)	登记号 (Cat. No.)	腹壳(ventral valve)		背壳(dorsal valve)	
		长(length)	宽(width)	长(length)	宽(width)
WH3	HB291	22	29		
DH2	HB311	19	28		
WH2	HB386	21	28		
DH2	HB92			约 23	约 32
WH2	Hb59				

腹内:铰齿细脊状;齿板发育,沿着腹肌痕面向前延伸,但其前端不会合;腹肌痕面大,近长卵形;启肌痕面大,月牙状,位于闭肌痕面两外侧;闭肌痕面窄小,位于整个肌痕面纵中部,但界线往往不很清楚。

背内:铰窝呈细坑状;内铰窝脊刀刃状,强烈异向展伸(图版 6,图 4b);背窗台宽阔;主突起强大,三叶型,两侧主突起极粗壮,椭圆形,内部发育 4~5 片主突起叶;中突起相对较小,呈方柱形,其内无主突起叶,但中突起后方具一显著中沟(图版 6,图 5,8);背窗台与其前轻微隆起的背壳底融合形成强大、三角状背台；背台前被3根近于平行的短隔脊所支持(图版6,图4b-4c)；背肌痕面位于背台内,但不易区分(图版6,图4b-4c,7)。

比较：*Leptaenopoma yichangense* 与 *L. trftidum* 以及 *L. rugosas* 的主要不同是：*L. yichangense* 两侧主突起极为粗壮,呈椭圆形,其前端远长于中突起,在其内部还具有4～5片主突起叶(图版6,图5,6b,9)；而其中突起则短小,呈方柱形,在其后方具一显著中沟(图版6,图5,8；插图12-C)；上述这些特征与后2个种有着明显的差别。

产地层位：湖北宜昌黄花场、丁家坡、王家湾；上奥陶统顶部五峰组观音桥段(赫南特阶中部)。

雕月贝科 Family Glyptomenidae Williams,1965

雕月贝亚科 Subfamily Glyptomeninae Williams,1965

平月贝属 Genus *Paromalomena* Rong,1979

1979　*Paromalomena* Rong. 6 页。
1980　*Bracteoleptaena* Havlicek(1967);Nikitin. P. 59.
1981　*Paromalomena* Rong;Chang. 562 页。
1983　*Paromalomena* Rong;Zeng. 119 页。
1984　*Paromalomena* Rong;Rong. P. 150.
2000　*Paromalomena* Rong;Cocks et Rong. P. 254.
2006　*Paromalomena* Rong;Rong;297 页。

属型种：Genotype *Platymena*? *polonica* Temple,1965.

特征简述：贝体小,轮廓横亚长方形或横半圆形；侧视平凸型,贝体很薄；铰合线直,稍微短于最大壳宽；主端近直角状或钝角状。腹壳凸度低缓；腹铰合面低；腹三角孔后部覆有假三角板。背壳凸度很低,壳面平坦；背铰合面很低；背三角孔被背三角板覆盖。壳表饰微弱、细密放射纹；同心层或同心皱微弱,但有时较显著。假疹壳。

腹内：齿板薄板状,异向展伸；在腹窗腔后端具一根圆柱形肉茎管,并穿越假三角板后端；腹肌痕模糊。

背内：铰窝显著,长三角状；内铰窝脊薄板状,强烈异向展伸；背窗台浅小,位于铰合线之后；主突起双叶型,短小,其两前侧有时与内铰窝脊后端互相融合,有时在主突起前方具一深宽主突起坑；背肌痕面模糊。

分布及时代：世界各地；晚奥陶世赫南特期中期(Middle Hirnantian)。

波兰平月贝 *Paromalomena polonica* (Temple)

图版(pl.)14,图(figs.)5-11;图版(pl.)15,图(figs.)1-12

1965　*Platymena*? *polonica* Temple,P. 407-410,pl. 15,figs. 1-4;pl. 16,figs. 1-5.
1967　*Bracteoleptaena polonica*(Temple);Marek et Havliček. P. 283,pl. 4,figs. 3,6,9.
1967　*Bracteoleptaena polonica*(Temple);Havliček. P. 113,pl. 20,figs. 18-22.
1975　*Bracteoleptaena polonica*(Temple);Fu. P. 113,pl. 24,figs. 1-4.
1978　*Bracteoleptaena polonica*(Temple);Yan. P. 223,pl. 62,figs. 16,17.
1979　*Paromalomena polonica*(Temple);Rong. 2 页,图版2,图2-4。
1980　*Bracteoleptaena polonica*(Temple)Nikitin. P. 59-61,pl. 15,figs. 1-10,12.
1981　*Paromalomena polonica* (Temple);Chang. 562 页,图版1,图29,34-36。
1983　*Paromalomena polonica* (Temple);Zeng. 119 页,图版17,图7-10。
1984　*Paromalomena polonica* (Temple);Rong. P. 294,pl. 2,figs. 3,4,7.

描述：壳小,通常壳长3.2～6.5mm,壳宽5～10.5mm(表6)；轮廓近横亚长方形或近半圆形；侧视平凸型,贝体很薄;铰合线直,近等于最大壳宽;主端近直角或钝角状。腹壳凸度平缓;腹铰合面低;腹三角孔被假三角板覆盖。背壳平坦,仅在喙部附近轻微隆起;背铰合面很低;背三角孔被背三角板覆盖。壳表饰微弱放射纹;同心层或同心皱微弱,但有时较显著。假疹壳。

表6 波兰平月贝介壳测量(单位:mm)
Table 6 Shell measurements of *Paromalomena polonica* (Temple) (in mm)

采集号 (Coll. No.)	登记号 (Cat. No.)	腹壳(ventral valve)		背壳(dorsal valve)	
		长(length)	宽(width)	长(length)	宽(width)
WH2	HB444	3.8	5.1		
WH2	HB394			2	3.1
WH3	HB467			6.5	10.5
WH3	HB457	13.5	20		
WH2	HB161	4.8	6.2		
WH3	HB456	7	10		
WH3	HB460	3.2	5		
WH3	HB586	6.2	9.6		
WH1	HB446			4.9	8
WH2	HB469	3.1	4		
WH2	HB445			3.2	4.5

腹内:齿板薄,约呈125°角异向展伸;腹窗腔后端具一根圆柱状肉茎管,并穿越假三角板后端伸出喙部(图版14,图5;图版15,图2,4,5);腹肌痕模糊,但有时隐约可见,有时具3根细隔板(图版15,图6)。

背内:铰窝小,长三角形;内铰窝脊薄板状,约呈140°角异向展伸;背窗台浅小,位于铰合线之后;双叶主突起短小,有时其两前侧与内铰窝脊后端融合;有时主突起之前具一深宽主突起坑(图版15,图9);背肌痕面模糊。

产地层位:湖北宜昌王家湾、丁家坡、黄花场;上奥陶统顶部五峰组观音桥段(赫南特阶中部)。

中华月贝科(新科) Family Sinomenidae Zeng, Chen et Zhang (fam. nov.)

模式属:Type genus *Sinomena* Zeng, Chen et Zhang (gen. nov.).

词源:建立在 *Sinomena* (gen. nov.)模式属基础上。

特征:贝体中等大,轮廓近半圆形,侧视平凸型或轻微凹凸型;齿板短宽,其表面不同部位具不同形态的齿板小牙饰(dental plate denticulates);内铰窝脊粗宽,具有不同数量的内铰窝脊小齿状突起(denticles on inner socket ridge),而铰合线上都是光滑的;主突起双叶型或三叶型;有的在主突起前方具主突起坑;有的具短背中隔脊。

讨论:Sinomenidae(fam. nov.)各成员的齿板短宽,具不同形态的齿板小牙饰;内铰窝脊粗宽,具有不同数量的内铰窝脊小齿状突起;铰合线光滑。这一独特的族群与在 Strophomenoidea King(1846)超科内的 Amphistrophiidae Harper(1973), Douvillinidae Caster(1939), Leptostrophiidae Caster(1939), Eopholidostrophiidae Rong et Cocks (1994), Strophodontidae Caster (1939), Shaleriidae Williams (1965)和 Strophonellidae Coster(1939)等各个科的成员都为铰合线小牙饰(hinge line denticulate) (Coks et Rong, 2000)有着本质的差别。因此以 *Sinomena* (gen. nov.)为模式属在 Strophomenoidea King(1846)超科内建立 Sinomenidae (fam. nov.)这一新科。目前这个新科 Sinomenidae (fam. nov.)包含 *Aphanomena* Bergström(1968), *Eostropheodonta* Bancroft(1949), *Yichangomena* (gen. nov.), *Sinomena*(gen. nov.)和 *Hubeinomena*(gen. nov.)5个属。

时代：晚奥陶世赫南特期中期至志留纪兰多维列世。

隐月贝属 Genus *Aphanomena* Bergström,1968

1968 *Aphanomena* Bergström, P. 13.
1978 *Aphanomena* Bergström; Yan. 223 页。
1979 *Aphanomena* Bergström; Rong. 2 页。
1980 *Rafinesquina* Hall et Clarke; Nikitin. P. 54.
1983 *Aphanomena* Bergström; Zeng. 120 页。
1984 *Aphanomena* Bergström; Rong. P. 156.
2000 *Eostropheodonta* Bancroft; Cocks et Rong, P. 282.

属型种：Genotype *Aphanomena schmalenseei* Bergström,1968.

特征简述：贝体中等大，轮廓近半圆形，侧视平凸或轻微凹凸型；腹三角孔被假三角板覆盖；背三角孔被背三角板覆盖。齿板短粗，异向展伸，其中部表面具数根与齿板侧缘近平行的齿板小牙饰（dental plate denticulates）。三叶状主突起；中主突起小，细球状，位于主突起冠部后端；两侧主突起强壮，窄距离异向展伸至背窗腔前缘；内铰窝脊粗宽，其表面各具3～6根近垂直于内铰窝脊侧边的内铰窝脊小齿状突起（denticles on inner socket ridge）。背、腹肌痕面不清晰，无背、腹中隔板。壳纹细，作2～3次分枝或插入式增多；同心生长纹微弱。假疹壳。

讨论：*Aphanomena* Bergström(1968)与*Eostropheodonta* Bancroft(1949)这两个属的外形和壳表特征的确很相似，加上两个属的正型标本保存不太好，因此导致*Aphanomena*被视为*Eostropheodonta*的同义名(Cocks et Rong,2000,P. 282)。但是，当前我们在与该两属正型标本产出层位相当的宜昌地区五峰组观音桥段（观音桥层）采集到许多较好的背、腹内模标本，无论是齿板小牙饰发生的部位还是主基形态，这两个属的性质都有明显的区别。*Aphanomena*的齿板小牙饰是发生在齿板中部(Rong,1984, P. 157,fig. 15；本书图版9，图4b；插图13-A)；主突起三叶型，内铰窝脊粗宽，各具6根近垂直于内铰窝脊侧边的内铰窝脊小齿状突起（图版9，图7a,7b；插图13-B）。而*Eostropheodonta*齿板小牙饰是发生在齿板后边缘（图版8，图2b；插图14-A），主突起为双叶型（图版7，图7-9；插图14-B）。因此，*Aphanomena*与*Eostropheodonta*是有着本质的不同，应单独为一个属，而不能作为*Eostropheodonta*的同义名。*Aphanomena*目前仅包含*A. schmalenseei* Bergström(1968)，*A. ultrix* (Marek et Havlíček, 1967)，*A. parvicostellata* Rong(1984)3个种。

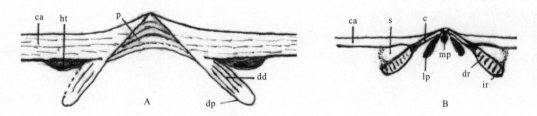

插图13　*Aphanomena parvicostellata* Rong 的内部构造

Illustr. 13　Interior structures of *Aphanomena parvicostellata* Rong

A-腹内膜（据图版9，图4b）；B-背内模（据图版9，图7a）

A - Ventral internal mold(from pl. 9, fig. 4b); B - Dorsal internal mold(from pl. 9, fig. 7a)

c-背三角板(chilidium); ca-铰合面(cardinal area); dd-齿板小牙饰(dental plate denticulates); dp-齿板(dental plate); dr-内铰窝脊小齿状突起(denticles on inner socket ridge); ht-铰齿(hinge tooth); ir-内铰窝脊(inner socket ridge); lp-侧主突起(lateral cardinal process); mp-中主突起(median cardinal process); p-假三角板(pseudodeltidium); s-铰窝(socket)

分布及时代：中国中南部、中欧和西欧；晚奥陶世赫南特期中期。

微壳纹隐月贝 *Aphanomena parvicostellata* Rong,1984

图版(pl.)9,图(figs.)1-8;图版(pl.)10,图(figs.)1-10;
图版(pl.)51,图(fig.)1

1977 *Eostropheodonta ultrix*(Marek et Havlíček);Zeng.61 页,图版 21,图 7,8。
1979 *Aphanomena ultrix*(Marek et Havlíček),Rong,P.2,pl.1,figs.19,20。
1983 *Aphanomena ultrix*(Marek et Havlíček,1967),Zeng.120 页,图版 17,图 1-3。
1984 *Aphanomena parvicostellata* Rong,P.156-159,pl.13,figs.1-10,15。
2006 *Eostropheodonta parvicostellata*(Rong,1984),Rong,294 页。

描述:贝体中等大,通常壳长 9～23mm,壳宽 12～28mm(表 7);轮廓近半圆形,侧缘和前缘近圆弧状;侧视平凸型至轻微凹凸型;铰合线直,近等于最大壳宽;主端近直角状。腹壳凸度低缓,仅在喙部前方稍微隆起;腹铰合面低,轻微斜倾型;腹三角孔被假三角板覆盖。背壳凸度平坦或者轻微凹下;背铰合面很低,正倾型;背三角孔被背三角板覆盖(插图 13-B)。壳表饰微型壳纹,壳纹细圆,通常作 2～3 次分叉或插入式增加;同心生长纹微弱,密集。假疹壳。

表 7 微壳纹隐月贝介壳测量(单位:mm)
Table 7 Shell measurements of *Aphanomena parvicostellata* Rong(in mm)

采集号 (Coll. No.)	登记号 (Cat. No.)	腹壳(ventral valve)		背壳(dorsal valve)	
		长(length)	宽(width)	长(length)	宽(width)
WH2	HB577	20.2	25		
WH2	HB589	23	28		
WH2	HB44			9	12.3
WH1	HB488			14.8	18.2

腹内:铰齿细脊状;齿板短粗,异向展伸;齿板后沟较长,平直,与铰合线方向一致(pl.9,fig.4b);在齿板表面中部具 3～4 根、近平行于齿板侧边的齿板小牙饰(dental plate denticulates)(图版 9,图 4b;插图 13-A);腹肌痕面模糊不清,但有的可见,腹肌痕面较小,闭肌痕中间具一纵中沟,将腹闭肌痕分成两个长条状;启肌痕位于闭肌痕两外侧,较大,但较模糊。

背内:铰窝长三角状;内铰窝脊粗宽,异向展伸,其表面具一排显著、4～6 根近垂直于内铰窝脊侧边的内铰窝脊小齿状突起(denticles on inner socket ridge)(图版 9,图 7a,7b;插图 13-B);主突起三叶型;中主突起小,圆突状,位于主突起冠部的后方(图版 51,图 1);两侧主突起强壮,轻微异向展伸,伸达背窗腔前缘;背肌痕面模糊。

比较:*Aphanomena parvicostellata* Rong(1984)与 *A. schmalenseei* Bergström(1968)的主要区别是 *A. parvicostellata* 的贝体较小,壳表放射纹更为细密,内铰窝脊小齿状突起较为粗壮;而 *A. schmalenseei* 的贝体大(壳宽达 50mm 以上),壳线粗宽。

产地层位:湖北宜昌王家湾、丁家坡;上奥陶统顶部五峰组观音桥段(赫南特阶中部)。

始齿扭贝属 Genus *Eostropheodonta* Bancroft,1949

1949 *Eostropheodonta* Bancroft.P.9.
1965 *Eostropheodonta* Bancroft;Williams.H395.
1965 *Eostropheodonta* Bancroft;Temple.P.410.
2000 *Eostropheodonta* Bancroft;Cocks et Rong.P.282.

属型种:Genotype *Orthis hirnantensis* M'Coy,1851。

特征简要：贝体中等大，轮廓近半圆形，侧视平凸型或轻微凹凸型；铰合线直，近等于最大壳宽；主端近直角状。腹壳缓凸，腹三角孔被假三角板覆盖。背壳平坦或微凹；背三角孔被背三角板覆盖。壳表饰微型壳纹或簇型壳纹；同心生长纹微弱。假疹壳。

腹内：齿板短粗，异向展伸；齿板后沟长，并与铰合线方向一致，限制着齿板小牙饰（图版50，图1；插图14-A）；齿板后缘具一排7～10根近垂直于齿板后沟的齿板小牙饰（dental plate denticulates）（图版8，图2b）（这与Williams，1965，H396，fig. 255-3e；Nikitin，1980，pl. 17，fig. 15b，所识别的 *Eostropheodonta* 的齿板小牙饰的特征相同）；腹肌痕面模糊。

背内：铰窝长三角状；内铰窝脊强壮，异向展伸，其表面具数个近垂直于内铰窝脊侧边的内铰窝脊小齿状突起（denticles on inner socket ridge）；主突起双叶型；背肌痕面不清晰。

比较：*Eostropheodonta* Bancroft（1949）与 *Aphanomena* Bergström（1968）的贝体轮廓和壳饰特征都很相似。但是 *Eostropheodonta* 的齿板小牙饰是发生在齿板的后缘，主突起为双叶型；而 *Aphanomena* 的齿板小牙饰是产生在齿板中部，主突起为三叶型，因此该两属有本质的差别，也因此不能把 *Aphanomena* 作为 *Eostropheodonta* 的同义名。

分布及时代：中国中南部，中欧和西欧；晚奥陶世赫南特期中期。

赫南特始齿扭贝 *Eostropheodonta hirnantensis*（M′Coy）

图版（pl.）7，图（figs.）1-12；图版（pl.）8，图（figs.）1-11；

图版（pl.）50，图（figs.）1，2

1949	*Eostropheodonta hirnantensis*（M′Coy，1851）；Bancroft. P. 9.
1965	*Eostropheodonta hirnantensis*（M′Coy）；Temple. P. 410-412，pl. 17，figs. 1-6；pl. 18，figs. 1-7；pl. 19，figs. 1-5.
1967	*Rafinesquina ultrix* Marek et Havliček. P. 282，pl. 3，figs. 1-3，6，8.
1967	*Rafinesquina urbicola* Marek et Havliček. P. 281-282，pl. 3，figs. 4，5，7.
1977	*Eostropheodonta ultrix*（Marek et Havliček）Zeng. 61页，图版21，图7，8.
1983	*Aphanomena rugosa* Zeng. 120页，图版17，图11。
2006	*Eostropheodonta parvicostellata*（Rong）；Rong. 297页，图版2，图17，21.

描述：贝体中等大，成年贝体通常壳长13～22mm，壳宽18～31mm（表8）；轮廓近半圆形；铰合线直，稍微短或近等于最大壳宽；主端近直角状或钝圆状；侧视平凸或轻微凹凸型，体腔区的贝体很薄。腹壳缓凸，仅在喙部前方稍微隆起；腹铰合面低，轻微斜倾型；腹三角孔被假三角板覆盖。背壳平坦或轻微凹下；背铰合面很低，正倾型；背三角孔被背三角板覆盖。壳表饰放射纹，作2～3次分枝或插入式增加，但有时壳纹重叠呈簇型放射纹。假疹壳。

表8 赫南特始齿扭贝介壳测量（单位：mm）
Table 8 Shell measurements of *Eostropheodonta hirnantensis*（M′Coy）（in mm）

采集号 (Coll. No.)	登记号 (Cat. No.)	腹壳（ventral valve）		背壳（dorsal valve）	
		长（length）	宽（width）	长（length）	宽（width）
DH2	HB346	15.8	19.5		
WH2	HB384	19.5	24		
DH2	HB536	13	18		
WH2	HB593			20	26
WH2	HB578			7.2	10
WH2	HB594			22	31

腹内：齿板短粗，异向展伸；在齿板后缘具一后沟，方向与铰合线一致；在齿板后沟前缘具一排7～10根近垂直于后沟的齿板小牙饰（图版50，图1；插图14-A）；腹肌痕面不清晰。

背内：铰窝三角状；内铰窝脊显著，异向展伸，其表面具3～4根近垂直于内铰窝脊侧边的内铰窝脊小齿状突起；主突起双叶型；背肌痕面模糊。

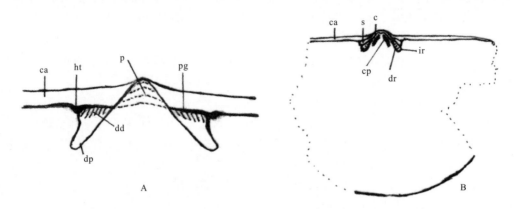

插图14 *Eostropheodonta hirnantensis*（M'Coy）的内部构造
Illustr. 14 Interior structures of *Eostropheodonta hirnantensis*（M'Coy）
A-腹内模（据图版8，图2b）；B-背内模（据图版8，图9）
A - Ventral internal mold（from pl. 8, fig. 2b）；B - Dorsal internal mold（from pl. 8, fig. 9）
c-背三角板（chilidium）；ca-铰合面（cardinal area）；cp-双叶状主突起（bilobed cardinal process）；dd-齿板小牙饰（dental plate denticulates）；dp-齿板（dental plate）；dr-内铰窝脊小齿状突起（denticles on inner socket ridge）；ht-铰齿（hinge tooth）；ir-内铰窝脊（inner socket ridge）；p-假三角板（pseudodeltidium）；pg-后沟（posterior groove）；s-铰窝（socket）

讨论：从当前标本的轮廓呈半圆形，齿板后沟前缘具有齿板小牙饰，内铰窝脊具一排内铰窝脊小齿状突起，主突起双叶型，壳表为密型壳纹或簇型壳纹来看，这些标本应为 *Eostropheodonta hirnantensis*（M'Coy），并与 Temple(1965) *E. hirnantensis*（M'Coy）(pls. 17-19) 雷同。

产地层位：湖北宜昌王家湾、丁家坡；上奥陶统顶部五峰组观音桥段（赫南特阶中部）。

宜昌月贝属（新属）Genus *Yichangomena* Zeng, Zhang et Han(gen. nov.)

属型种：Genotype *Yichangomena dingjiapoensis* Zeng, Zhang et Han(gen. et. sp. nov)。

词源：Yichang 为化石产地的汉语拼音，Mene（希腊语），新月状，表示新属壳形如半月状。

特征简述：轮廓近半圆形；铰合线直，近等于最大壳宽；侧视平凸型；壳表饰密型壳纹；假疹壳。齿板后沟长，向腹三角孔后侧边斜伸；齿板小牙饰（dental plate denticulates）发生在齿板并向后扩展至腹三角孔两后侧边，并与腹三角孔侧边近平行。主突起三叶型，但中突起微小；主突起坑深宽，位于主突起前方，限制着主突起；内铰窝脊小齿状突起（denticles on inner socket ridge）3～5根。

描述：贝体小至中等大，轮廓近半圆形；铰合线直，近等于最大壳宽；主端近直角状；侧视平凸或轻微凹凸型，贝体很薄。腹壳凸度低缓，仅在喙部前方稍微隆起；腹铰合面低，轻微斜倾型；腹三角孔被假三角板覆盖。背壳平坦或轻微凹下；背铰合面很低，正倾型；背三角孔被背三角板覆盖。壳表饰放射纹，作2～3次插入式增加，同心生长纹微弱。假疹壳。

腹内：齿板短粗，异向展伸；齿板后沟较长，向腹三角孔后侧边斜伸（图版52，图1）；具一排15～16根的齿板小牙饰，它们发生在齿板表面并向后扩展至腹三角孔两后侧边，而且与腹三角孔侧边近于平行（图版52，图1,2；插图15-A）；腹肌痕面不清晰。

背内：铰窝小，近三角形；内铰窝脊显著，异向展伸；在内铰窝脊表面具一排3～5根近垂直于内铰窝

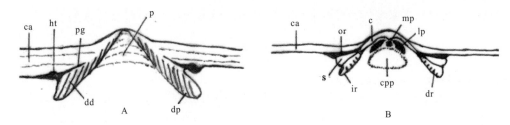

插图 15 *Yichangomena dingjiapoensis*(gen. et sp. nov.)的内部构造

Illustr. 15 Interior structures of *Yichangomena dingjiapoensis*(gen. et sp. nov.)

A-腹内模(据图版 11,图 8);B-背内模(据图版 11,图 4a,4b)

A - Ventral internal mold(from pl. 11, fig. 8);B - Dorsal internal mold(from pl. 11, figs. 4a,4b)

c-背三角板(chilidium);ca-铰合面(cardinal area);cpp-主突起坑(cardinal process pit);dd-齿板小牙饰 (dental plate denticulates);dp-齿板(dental plate);dr-内铰窝脊小齿状突起(denticles on inner socket ridge);ht-铰齿(hinge tooth);ir-内铰窝脊(inner socket ridge);lp-侧主突起(lateral cardinal process); mp-中主突起(median cardinal process);or-外铰窝脊(outer socket ridge);p-假三角板(pseudodeltidium); pg-后沟(posterior groove);s-铰窝(socket)

脊侧边的内铰窝脊小齿状突起,(图版 11;图 4b;插图 15 - B);主突起三叶型,但中主突起微小;主突起坑深宽,位于主突起前方,限制着主突起;背肌痕面模糊。

比较:新属 *Yichangomena*(gen. nov.)与 *Aphanomena* Bergström(1968)的贝体轮廓和外部特征都很相似。但是 *Yichangomena*(gen. nov.)的贝体较小,后沟向后斜伸,齿板小牙饰布满整个齿板和腹三角孔两侧边,主突起坑深宽。而 *Aphanomena* 的贝体较大,后沟平伸并与铰合线方向一致,齿板小牙饰只发生在齿板中部,三叶状主突起之前没有主突起坑。因此,该两属有本质上的差别。

分布及时代:中国中南部;晚奥陶世赫南特期中期。

丁家坡宜昌月贝(新属、新种)*Yichangomena dingjiapoensis* Zeng, Zhang et Han(gen. et sp. nov.)

图版(pl.)11,图(figs.)1-11;图版(pl.)52,图(figs.)1,2

词源:Dingjiapo 为化石产地的汉语拼音;ensis 为化石种名常用的词尾,表示来源。

描述:贝体小至中等,通常壳长 5.5~10mm,壳宽 6~13.5mm(表 9);轮廓近半圆形;侧视平凸或轻微凹凸型;铰合线直,近等于最大壳宽;主端近直角状;侧缘和前缘近于圆弧状。腹壳凸度低缓,仅在喙部前方稍微隆起;腹铰合面低,斜倾型;腹三角孔被假三角板覆盖。背壳平坦或轻微凹下;背铰合面很低,正倾型;背三角孔被背三角板覆盖。壳表饰放射纹,作 2~3 次插入式增加;同心生长纹极微弱。假疹壳。

背、腹内部构造特征同属征描述。

表 9 丁家坡宜昌月贝介壳测量(单位:mm)

Table 9 Shell measurements of *Yichangomena dingjiapoensis*(gen. et sp. nov.)(in mm)

采集号 (Coll. No.)	登记号 (Cat. No.)	腹壳(ventral valve)		背壳(dorsal valve)		备注 (remarks)
		长(length)	宽(width)	长(length)	宽(width)	
WH1	HB100	5.5	6			
DH2	HB345	6	7.2			副型(paratype)
DH2	HB519			5.5	6.9	正型(holotype)
WH3	HB568			10	13.5	副型(paratype)

产地层位：湖北宜昌丁家坡、王家湾；上奥陶统顶部五峰组观音桥段（赫南特阶中部）。

中华月贝属（新属） Genus *Sinomena* Zeng, Chen et Zhang (gen. nov.)

属型种：Genotype *Sinomena typica* Zeng, Chen et Zhang (gen. et sp. nov.).

词源：Sino（英文），表示中国的；mena 为扭月贝类属名的词尾，表示新属的形状如半月形。

特征简述：齿板具一排近垂直于齿板内侧边的齿板小牙饰（dental plate denticulates）；内铰窝脊具一排近垂直于内铰窝脊侧边的内铰窝脊小齿状突起（denticles on inner socket ridge）；三叶型主突起之前具一短背中隔脊。

描述：贝体中等大，轮廓近半圆形；侧视平凸或轻微凹凸型；铰合线直，近等于最大壳宽；主端近直角状；侧缘和前缘近圆弧状。腹壳凸度低，仅在喙部前方稍微隆起；腹铰合面低，斜倾型；腹三角孔被假三角板覆盖。背壳平坦或轻微凹下；背铰合面很低，正倾型；背三角孔被背三角板覆盖。壳表饰放射纹，作 2～3 次插入式或分枝式增多；同心纹细密，微弱。假疹壳。

腹内：齿板短粗，异展伸；齿板后沟较短，平直，与铰合线方向基本一致；在齿板表面具一排 4～6 根近垂直于齿板内侧边的齿板小牙饰（图版 53，图 1；插图 16 - A）；腹肌痕面不清晰。

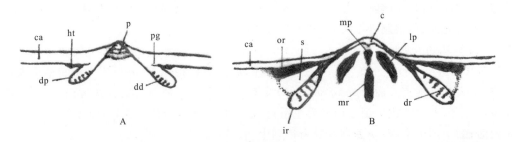

插图 16　*Sinomena typica*（gen. et sp. nov.）的内部构造

Illustr. 16　Interior structures of *Sinomena typica*（gen. et sp. nov.）

A-腹内模（据图版 12，图 8）；B-背内模（据图版 12，图 3b - 3c）

A – Ventral internal mold (from pl. 12, fig. 8); B – Dorsal internal mold (from pl. 12, figs. 3b – 3c)

c-背三角板（chilidium）；ca-铰合面（cardinal area）；dd-齿板小牙饰（dental plate denticulates）；dp-齿板（dental plate）；dr-内铰窝脊小齿状突起（denticles on inner socket ridge）；ht-铰齿（hinge tooth）；ir-内铰窝脊（inner socket ridge）；lp-侧主突起（lateral cardinal process）；mp-中主突起（median cardinal process）；mr-中隔脊（median ridge）；or-外铰窝脊（outer socket ridge）；p-假三角板（pseudodeltidium）；pg-后沟（posterior groove）；s-铰窝（socket）

背内：铰窝显著，长三角状；外铰窝脊显著；内铰窝脊粗壮，异向展伸，其表面具一排 4～5 根近垂直于内铰窝脊侧边的内铰窝脊小齿状突起，（图版 12，图 3c；插图 16 - B）；主突起三叶型，两侧的主突起强壮，小角度异向伸达至背窗腔前缘；中突起微小，但向腹方耸立；背中隔脊短，其前端向腹方翘起（图版 12，图 3b），而其后端始于中突起前缘（图版 53，图 2）或始于中突起前方（图版 12，图 3b，4b）。

比较：*Sinomena*（gen. nov.）与 *Aphanomena* Bergström（1968）的主要区别是 *Sinomena*（gen. nov.）的背内具短粗背中隔脊，腹内具一排垂直于齿板内侧边的齿板小牙饰；而 *Aphanomena* 的背内没有背中隔脊，腹内的齿板小牙饰发生在齿板中部，并与齿板内侧边近于平行。*Sinomena*（gen. nov.）与 *Yichangomena*（gen. nov.）的主要区别是 *Sinomena*（gen. nov.）的三叶型主突起之前具背中隔脊，齿板小牙饰近垂直于齿板内侧边；而 *Yichangomena*（gen. nov.）的三叶型主突起之前具一宽深的主突起坑，齿板小牙饰不仅布满整个齿板，而且还向后扩展到腹三角孔两后侧边，并且与腹三角孔两侧边近于平行。

分布及时代：中国中南部；晚奥陶世赫南特期中期。

标准中华月贝(新属、新种)*Sinomena typica* Zeng, Chen et Zhang (gen. et sp. nov.)

图版(pl.)12,图(figs.)1-9;图版(pl.)53,图(figs.)1,2

词源：Typica(拉丁文),典型的,表示为本新属的典型代表。

描述：贝体中等大,通常壳长 11.5～18mm,壳宽 16.2～23mm(表 10);轮廓近半圆形;侧视平凸或轻微凹凸型;铰合线直,近等于最大壳宽;主端近直角状;侧缘和前缘呈圆弧状。腹壳凸度低缓,仅在喙部前方稍微隆起;腹铰合面低,斜倾型;腹三角孔被假三角板覆盖。背壳平坦或轻微凹下;背铰合面很低,正倾型;背三角孔被背三角板覆盖。壳表饰放射纹,通常作 2～3 次插入式或分枝式增加;同心生长纹细密,而且很微弱。假疹壳。

表 10 标准中华月贝介壳测量(单位:mm)
Table 10 Shell measurements of *Sinomena typica* (gen. et sp. nov.) (in mm)

采集号 (Coll. No.)	登记号 (Cat. No.)	腹壳(ventral valve)		背壳(dorsal valve)		备注 (remarks)
		长(length)	宽(width)	长(length)	宽(width)	
WH3	HB588	5.4	6.5			
WH2	HB606	12	13			副型(paratype)
WH2	HB564			18	23	副型(paratype)
WH2	HB563			11.5	16.2	正型(holotype)

背、腹内部构造同属征。

产地层位：湖北宜昌王家湾;上奥陶统顶部五峰组观音桥段(赫南特阶中部)。

湖北月贝属(新属)Genus *Hubeinomena* Zeng, Chen et Zhang (gen. nov.)

属型种：Genotype *Hubeinomena wangjiawanensis* Zeng, Chen et Zhang (gen. et sp. nov.).

词源：Hubei 为化石产地所在省份的汉语拼音;omena 为扭月贝类属名的词尾。

特征简述：贝体中等大,轮廓半圆形;侧视平凸或轻微凹凸型;齿板短宽,异向展伸;齿板前部具 2 根粗壮小牙饰(dental plate denticulates),并与齿板侧边轻微斜交;主基开阔,但很短;主突起密集双叶型,短而粗壮;内铰窝脊具 2～3 个小齿状突起(denticles on inner socket ridge)。壳线较简单,线顶圆滑,2 次分叉,同心生长纹极为细密。

描述：贝体中等大,体腔区狭窄;轮廓半圆形;侧视平凸型或轻微凹凸型;铰合线直,近等于最大壳宽;主端近直角状或近钝角状。腹壳凸度低缓,仅在顶区稍微隆起;腹铰合面低,斜倾型;腹喙大,不弯曲;腹三角孔被假三角板覆盖。背壳凸度低平或轻微凹;背铰合面很低;背喙很小;背三角孔被背三角板覆盖。壳表饰简单放射线,线顶圆滑,作 2 次分枝式或者插入式增多;同心生长纹细密。假疹壳。

腹内：铰齿小,细脊状;齿板短、宽,异向展伸;在齿板上具 2 根短粗小牙饰(dental plate denticulates),并与齿板侧边轻微斜交(图版 13,图 1b,1c;插图 17-A);腹肌痕面隐约可见,亚圆形;闭肌痕居中,近长方形;启肌痕稍微大些,位于闭肌痕两外侧,呈月牙状(图 14,图 2)。

背内：主基开阔,但很短;铰窝小,窄坑状;内铰窝脊宽,强烈异向展伸,在其表面具 2～3 个小齿状突起(denticles on inner socket ridge)(图版 13,图 5a-5c;图版 54,图 1;插图 17-B);主突起呈密集双叶型,短而粗壮(图版 13,图 3,5a,6);背肌痕面隐约可见,轻微隆起,呈倒瓶状(图版 13,图 5a;图版 54,图 2),但不易区分。

比较：新属 *Hubeinomena* (gen. nov.)与 *Eostropheodonta* Bancroft(1949)的外表和双叶型主突起有些相似。它们的主要区别是 *Hubeinomena* (gen. nov.)的放射线较简单,线顶圆滑,作 2 次分枝或插入式

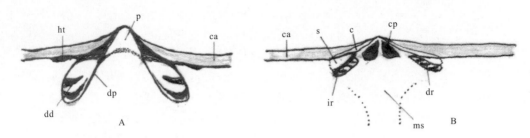

插图 17 *Hubeinomena wangjiawanensis*(gen. et sp. nov.)的内部构造

Illustr. 17 Interior structures of *Hubeinomena wangjiawanensis*(gen. et sp. nov.)

A-腹内模(据图版 13,图 1b,1c);B-背内模(据图版 13,图 5a-5c;图版 54,图 1-3)

A - Ventral internal mold(from pl. 13,figs. 1b,1c);B - Dorsal internal mold(from pl. 13,figs. 5a-5c;pl. 54,figs. 1-3)

c-背三角板(chilidium);ca-铰合面(cardinal area);cp-双叶状主突起(bilobed cardinal process);dd-齿板小牙饰(dental plate denticulates);dp-齿板(dental plate);dr-内铰窝脊小齿状突起(denticles on inner socket ridge);ht-铰齿(hinge tooth);ir-内铰窝脊(inner socket ridge);ms-肌痕(muscle scars);p-假三角板(pseudodeltidium);s-铰窝(socket)

增多;仅在齿板前部具 2 根粗壮的齿板小牙饰,而且与齿板侧边轻微斜交;主基开阔,但很短;双叶型主突起很短,很粗壮;在内铰脊上仅具 2~3 个粗壮的小齿状突起。

新属 *Hubeinomena*(gen. nov.)的贝体轮廓、壳表特征,以及齿板形态和双叶型主突起等情况与 *Fardenia* Lamont(1935)也很相似,其主要的区别是 *Fardenia* 的齿板上缺失小牙饰,内铰窝脊异向展伸角度较小,其表面无小齿状突起。

分布及时代:中国中南部;晚奥陶世赫南特期中期(Middle Hirnantian)。

王家湾湖北月贝(新属、新种)*Hubeinomena wangjiawanensis* Zeng, Chen et Zhang(gen. et sp. nov.)

图版(pl.)13,图(figs.)1-8;图版(pl.)14,图(figs.)1-4;
图版(pl.)54,图(figs.)1-3

词源:Wangjiawan 为化石产地的汉语拼音;ensis 为化石种名常用的词尾,表示来源。

描述:贝体中等大,通常壳长 21mm,壳宽 28mm,轮廓半圆形,贝体很薄;侧视平凸型或轻微凹凸型;铰合线直,近等于最大壳宽;主端近直角状或近钝角状。两侧缘和前缘近圆弧状。腹壳凸度低缓,仅在喙部前方稍微隆起;腹喙大,不弯曲;腹铰合面很低,斜倾型;腹三角孔被假三角板覆盖。背壳近平凸或轻微凹下;背喙很短小;背铰合面很低;背三角孔被背三角板覆盖。壳表饰简单放射线,线顶圆滑,一般作 2 次分枝,有时为插入式增加;同心生长纹极为细密(图版 14,图 1b)。

背、腹内构造同属征描述。

产地层位:湖北宜昌王家湾、黄花场;上奥陶统顶部五峰组观音桥段(赫南特阶中部)。

褶脊贝超科 Superfamily Plectambonitoidea Jones,1928

异脊贝科 Family Xenambonitidae Cooper,1956

埃月贝亚科 Subfamily Aegiromeninae Havlíček,1961

埃月贝属 Genus *Aegiromena* Havlíček,1961

1961 *Aegiromena* Havlíček,1961.

1980 *Aegiromena* Havlíček;Havlíček et Branisa. P. 38.

属型种:Genotype *Leptaena aquila* Barrande,1848.

讨论:Havliček(1961)建立 *Aegiromena* 属时,强调了该属的主要特征是肌台(bema)围脊十分低,并以此与 *Aegiria* Öpik(1933)的肌台围脊显著而高强相区别。因此 Rong et Yang(1981)认为上述的差异仅是量的变化,而非本质的不同(P.171);Cocks et Rong(1989)进一步将 *Aegiromena* 视为 *Aegiria* 的同义名(P.122);但是 Cocks et Rong(2000)又对 *Aegiromena* 的含义进行了修订,认为该属的肌台多变,有时在一些标本中缺失,背肌痕微弱(P.334),而 Zeng et al.(2013,手稿)则认为该属的肌台原始,围脊仅发育在肌台前缘,而且断断续续。尽管有上述多种见解,但目前仍然有些标本还是很难进行辨别。

当前我们在研究宜昌地区 *Hirnantia* 动物群时,发现一批背内模标本存在着双中隔板(图版19,图4-9),而这些标本的肌台特征又与 *Aegiromena* 的很接近。因此我们又仔细观察 *Aegiromena* 的属型种 *Leptaena aquila* Barrande(1848)的图影,即 Williams(1965)H382,fig.244-7b;Cocks et Rong(1989),P.121,fig.87;Cocks et Rong(2000) P.336,fig.218-1d;以及 Havliček(1967),*Aegiromena aquila*(Barrande),pl.3,figs.7,9;*Aegiromena aquila praecursor*(Havliček),pl.4,fig.1;*Aegiromena descendens*(Havliček),pl.4,figs.12,16;Cocks et Fortey(1997),*Aegiromena planissima*(Reed,1915),pl.2,figs.14,15;Rong(2006),*Aegiromena ultima* Marek et Havliček(1967),pl.2,fig.15 等背内模标本的图影。发现上述那些背内模标本均存在着2根背中隔板,尤其是 Havliček et Branisa(1980),pl.2,figs.2,5更为显著。表明 *Aegiromena* Havliček(1961)的背内除具有肌台外,同时还存在着双背中隔板这一重要特征。因此对 *Aegiromena* Havliček(1961)的属征作如下修订。

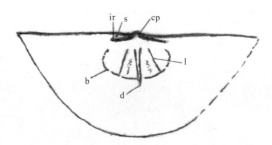

插图 18 *Aegiromena aquila*(Barrande)的背内模(据 Cocks et Rong,2000,fig.218-1d)

Illustr. 18 Dorsal internal mold of *Aegiromena aquila* (Barrande)(From Cocks et Rong,2000,fig.218-1d)
b-肌台(bema);cp-双叶状主突起(bilobed cardinal process);d-双背中隔板(diplosepta);l-侧隔板(lateral septum);ir-内铰窝脊(inner socket ridge);s-铰窝(socket)

修订后的特征简要:壳小,轮廓横半圆形;铰合线直,等于最大壳宽;主端尖翼状或近直角状;侧视平凸或轻微凹凸型。壳表饰放射纹,作2~3次插入式增多。假疹壳。

腹内:齿板短,宽距离异向展伸;腹肌痕面双叶状,中肌隔短粗。

背内:内铰窝脊宽距异向展伸;主突起双叶状,其前端与内铰窝脊后端互相融合;肌台发育程度不一,有强、有弱;具双背中隔板,其前端穿越肌台前缘(插图18)。

比较:*Aegiromena* Havliček(1961)与 *Aegiria* Öpik(1933a)的主要区别是 *Aegiromena* 的背内具双背中隔板,而 *Aegiria* 的背内则具单背中隔板。

分布及时代:中国中南部、泰国、波希米亚;中奥陶世晚期至晚奥陶世晚期。

双板埃月贝(新种) *Aegiromena diplosepta* Zeng,Peng et Zhang (sp. nov.)

图版(pl.)19;图(figs.)3-9;插图(Illustr.)19-A,B

1997 *Aegiromena planissima*(Reed,1915);Cocks et Fortey. pl.2,figs.14,15.
2006 *Aegiromena ultima* Marek et Havliček;Rong.305页,图版2,图15。

词源:Diplo(英文),双;Septa(英文)为 Septum 的复数,隔板;表示新种具有明显的双背中隔板。

描述:壳很小,通常壳长 3~3.2mm,壳宽 4.4~5.6mm(表11);轮廓半圆形;铰合线直,等于最大壳宽;主端近直角状;侧视平凸或轻微凹凸型。腹壳凸度平缓,仅在喙部前方稍微隆起;腹铰合面低,斜倾型;腹三角孔小,很可能被假三角板覆盖。背壳平坦或轻微凹下;背铰合面极低,呈线状;背三角孔可能被背三角板覆盖。壳表饰放射纹,一般作 2~3 次插入式增多。假疹壳。

腹内:齿板短小,异向展伸在腹肌痕面两后侧;腹肌痕面双叶状;启肌痕面较大,呈卵圆形;闭肌痕面很小，位于两个启肌痕之间的后部，被3根小肌隔分开（插图19-A）。

表 11 双板埃月贝（新种）介壳测量（单位：mm）
Table 11　Shell measurements of *Aegiromena diplosepta* (sp. nov.) (in mm)

采集号 (Coll. No.)	登记号 (Cat. No.)	腹壳(ventral valve)		背壳(dorsal valve)	
		长(length)	宽(width)	长(length)	宽(width)
WH2	HB36	3	4.4		
DH3	HB369			3.1	5
WH1	HB364			3	约4.4
DH2	HB358			3.2	5.6
WH1	HB365			3.2	4.8
DH2	HB361			3	4.5
DH3	HB367			3.3	约5.8

背内：铰窝小，三角状；内铰窝脊短宽，约呈145°角异向展伸；主突起小，双叶型，其两前端与内铰窝脊后端互相融合；主突起坑显著、近圆形，位于主突起前方；肌台微弱，但隐约可见，近横椭圆形；双背中隔板显著，从主突起坑前方开始伸出，直至穿越肌台前缘（图版19，图4-9；插图19-B）。

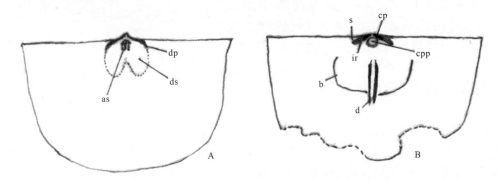

插图19　*Aegiromena diplosepta*（sp. nov.）的内部构造
Illustr. 19　Interior structures of *Aegiromena diplosepta*（sp. nov.）
A-腹内模（据图版19，图3）；B-背内模（据图版19，图4a,4b,7）
A – Ventral internal mold(from pl. 19, fig. 3); B – Dorsal internal mold(from pl. 19, figs. 4a, 4b, 7)
as-闭肌痕（adductor scars）；b-肌台（bema）；cp-双叶状主突起（bilobed cardinal process）；cpp-主突起坑（cardinal process pit）；d-双背中隔板（diplosepta）；dp-齿板（dental plate）；ds-启肌痕（diductor scars）；ir-内铰窝脊（inner socket ridg）；s-铰窝（socket）

比较：*Aegiromena diplosepta*（sp. nov.）的主要特征是主突起坑显著，具有双背中隔板，贝体轮廓呈半圆形，主端近直角状等情况与 *A. corolla* Havliček et Branisa(1980), *A. aquila*(Barrande, 1848), *A. aquila praecursor*(Havliček, 1952), *A. descendens* (Havliček, 1952)等种的主端呈尖翼状，或者贝体更横宽，或者肌台很显著，有着明显的不同。

产地层位：湖北宜昌丁家坡、王家湾；上奥陶统顶部五峰组观音桥段（赫南特阶中部）。

埃吉贝属 Genus *Aegiria* Öpik,1933

1933 *Aegiria* Öpik,P. 55.
1965 *Aegiria* Öpik;Williams. H381.
1967 *Aegiromena* Havliček;Marek et Havliček. P. 281.
1967 *Aegiromena* Havliček;Havliček. P. 45.
1970 *Aegiria* Öpik;Cocks. P. 195.
1977 *Aegiria* Öpik;Zeng. 59 页。
1979 *Aegiromena* Havliček;Rong. 9 页。
1980 *Aegiromena* Havliček;Nikitin. P. 52.
1983 *Aegiromena* Havliček;Zeng. 119 页。
1984 *Aegiromena* Havliček;Rong. P. 150.
1989 *Aegiria* Öpik;Cocks et Rong. P. 122.
1997 *Aegiromena* Havliček;Cocks et Fortey. P. 124.
2000 *Aegiria* Öpik;Cocks et Rong. P. 335.

属型种:Genotype *Aegiria norvegica* Öpik,1933.

特征简述:贝体小,轮廓半圆形;侧视轻微凹凸型;铰合线直,等于最大壳宽;主端锐角状或近直角状。腹壳凸度平缓,仅在喙部前方稍微隆起;腹铰合面低,斜倾型;腹三角孔被假三角板覆盖。背壳轻微凹下;背铰合面极低;背三角孔被背三角板覆盖。壳表饰放射纹,有时分为粗、细两组。假疹壳。

腹内:齿板短,展伸在腹肌痕面两后侧;腹肌痕面显著,双叶状;中肌隔显著,将腹肌痕面分成两部分。

背内:铰窝小,三角状;内铰窝脊短宽,强烈异向展伸;主突起小,双叶型,其前端与内铰窝脊互相融合;主突起坑显著,位于主突起前方;肌台多变,强弱不等,呈横椭圆形;背中隔板高强,从主突起坑之前伸出,其前端超越肌台前缘。

讨论:以往各研究者只强调 *Aegiria* Öpik(1933)具有高强的肌台(bema),并且以此和 *Aegiromena* Havliček(1961)为较弱的肌台相区别。但从目前大量标本证实并非如此。*Aegiromena* 的肌台有弱,也有强,而 *Aegiria* 的肌台也有强、弱的变化。目前经过我们仔细观察,*Aegiria* 与 *Aegiromena* 的主要区别是 *Aegiria* 仅具有单背中隔板,而 *Aegiromena* 则具有双背中隔板。

分布及时代:世界各地;晚奥陶世至早志留世(Llandoverian)。

平埃吉贝 *Aegiria planissima* (Reed,1915)

图版(pl.)16,图(figs.)1-12;图版(pl.)17,图(figs.)1-12

1915 *Schuchertella planissima* Reed. P. 78,pl. 11,figs. 13-19.
1967 *Aegiromena ultima* Marek et Havliček. P. 281,pl. 3,figs. 9-12.
1967 *Aegiromena ultima* Marek et Havliček;Havliček. P. 45,pl. 6,figs. 1-7.
1977 *Aegiromena ultima* Marek et Havliček;Zeng. 59 页,图版 20,图 11。
1979 *Aegiromena ultima* Marek et Havliček;Rong. 9 页,图版 1,图 21,22。
1980 *Aegiromena durbenensis* Nikitin. P. 53,pl. 13,figs. 16-22.
1983 *Aegiromena ultima* Marek et Havliček;Zeng. 119 页,图版 17,图 4-6。
1984 *Aegiromena ultima* Marek et Havliček;Rong. P. 150,pl. 11,figs. 10,13.
1984 *Aegiromena convexa* Chang;Rong. P. 148,pl. 11,figs. 2,4,5,9.
1997 *Aegiromena planissima* (Reed)Cocks et Fortey. P. 124,figs. 9-13,17,18.

描述:贝体很小,通常壳长 2.8~3mm,壳宽 4.7~5.3mm(表 12);轮廓半圆形,铰合线直,等于最大壳宽;主端锐角状或近直角状;侧视平凸或轻微凹凸型,贝体很薄。腹壳凸度平缓,仅在喙部前方稍微隆起;腹三角孔被小的假三角板覆盖;腹铰合面低;斜倾型。背壳平坦或轻微凹下;背铰合面很低;背三角孔被背三角板覆盖。

表12 平埃吉贝介壳测量(单位:mm)
Table 12　Shell measurements of *Aegiria planissima* (Reed, 1915) (in mm)

采集号 (Coll. No.)	登记号 (Cat. No.)	腹壳(ventral valve)		背壳(dorsal valve)	
		长(length)	宽(width)	长(length)	宽(width)
DH2	HB359	3.1	5.3		
WH1	HB414	2.8	4.8		
HK1	HB779	4	6.5		
DH2	HB343	2.8	4.7		
DH2	HB439	3	5		
DH3	HB143	4.2	5.3		
WH2	HB435	3	4.8		
DH2	HB330	3	4.5		
DH3	HB440	3	5.2		
DH2	HB427	3.8	5.3		
WH1	HB366	3.6	5.3		
DH2	HB434	3.2	4.9		
WH1	HB389	3	4.9		
DH2	HB423			3	4.7
DH2	HB415			3.3	5.2
HK3	HB780			3.2	5.2
DH2	HB428			3	4.7
WH2	HB380			2.9	
WH2	HB789			3.5	
DH2	HB426			2.9	4.8
DH2	HB327			2.9	5
DH2	HB429			2.9	4.7
DH2	HB552			3	4.9

腹内:齿板发育,向前延伸在腹肌痕面两外侧;腹肌痕面显著,双叶状;中肌隔短粗,将腹肌痕面一分为二(图版16,图3,5;图版17,图3,6)。闭肌痕很小,一对,呈倒尖三角状,居于肌痕面中后部;启肌痕很大,一对,呈椭圆形,位于闭肌痕两前侧,将闭肌痕夹在两启肌痕之间的中后部(图版17,图7)。

背内:铰窝小,三角状;内铰窝脊短宽,呈135°～140°角异向展伸;主突起短小,双叶型,其前端与内铰窝脊后端相融合(图版16,图8,12;图版17,图11);主突起前方具一个显著椭圆形主突起坑(图版17,图8-12);具锥形肌台,呈横椭圆形,其围脊很弱,时隐时现;背中隔板高强,始于主突起坑之前,向前穿越肌台前缘,伸达背壳底中部。

讨论:Cocks et Fortey(1997)在研究泰国奥陶纪末期 *Hirnantia* 动物群时指出:*Aegiromena ultima* Marek et Havlíček(1967)和 *Aegiromena convexa* Chang(Chang,1981;Rong,1984)都是 *Aegiromena* *planissima*(Reed,1915)的同义名。而当前所获得的大量背、腹内模标本(图版 16,图 1-12;图版 17,图 1-12)与 Cocks et Fortey(1997)从泰国所获得 *Aegiromena planissima* 的标本(pl. 2,figs. 9-13,17, 18)特征相同。但是,从前文对 *Aegiromena* Havliček(1961)的属征进行重新修订后得知:*Aegiromena* 的背内具有双背中隔板,因而它们不符合 *Aegiromena* 的特征,而恰与 *Aegiria* 的特征相符合,所以将它们改为 *Aegiria planissima*(Reed,1915)。

产地层位:湖北宜昌丁家坡、王家湾;上奥陶统顶部五峰组观音桥段(赫南特阶中部)。

似戟贝属 Genus *Chonetoidea* Jones,1928

1928	*Chonetoidea* Jones.	P. 393.
1953	*Sericoidea* Lindström.	P. 134.
1957	*Chonetoidea* Jones;Spjeldnaes.	P. 104.
1962	*Sericoidea* Lindström;Williams.	P. 187.
1965	*Chonetoidae* Jones;Williams.	H383.
1966	*Chonetoidea* Jones;Wang,Jin et Fang.	297 页.
1967	*Chonetoidea* Jones;Havliček.	P. 48.
1967	*Sericoidea* Lindström;Havliček.	P. 51.
1970	*Chonetoidea* Jones;Cocks.	P. 192.
1977	*Chonetoidea* Jones;Mitchell.	P. 93.
1981	*Sericoidea* Lindström;Chang.	562 页.
1982	*Sericoidea* Lindström;Fu.	118 页.
1987	*Chonetoidea* Jones;Zeng.	232 页.
1987	*Sericoidea* Lindström;Zeng.	232 页.
1989	*Chonetoidea* Jones;Cocks et Rong.	P. 123.
2000	*Chonetoidea* Jones;Cocks et Rong.	P. 337.

属型种:Genotype *Plectambonites papillosa* Reed(1905)。

特征简述:贝体小,轮廓横椭圆形或近半圆形,侧视平凸型或轻微凹凸型;铰合线直,等于最大壳宽;主端锐角状或近直角状。腹壳凸度平缓,仅在顶区稍微隆起;腹铰合面低,斜倾型;腹三角孔被小假三角板覆盖。背壳平坦或轻微浅凹;背铰合面极低;背三角孔可能被三角板覆盖。壳表饰 2 级壳纹;一级壳纹较粗,从喙部伸至前缘;二级壳纹较细,作 1~2 次插入于较粗壳纹之间;但有时为单一微细壳纹。假疹壳。

腹内:齿板短小,延伸在腹肌痕面两外侧;腹肌痕面双叶状显著,中间被短小肌隔分成两部分。

背内:铰窝小,三角状;内铰窝脊短宽,强烈异向展伸;主突起短小,双叶型,其两前端有时与内铰窝脊后端互相融合;主突起前方具一椭圆形主突起坑;肌台缺失;背中隔板薄而高,从主突起坑前方伸至背壳底中部;背肌痕面不清晰,但有时分为前、后两对。

背、腹壳底具规则或不规则、细长小突起。

讨论:由于 *Sericoidea* Lindström(1953,P. 134-139)与 *Chonetoidea* Jones(1928)极为近似,因此 Cocks et Rong,(1989,P. 123)将 *Sericoidea* 作为 *Chonetoidea* 同义名;加上 Jones(1928)建立 *Chonetoidea* 时的标本都保存不好,原属征定义不够明确,因此当前 *Chonetoidea* 的含义包含了原 *Sericoidea* 的一些主要特征。*Chonetoidea* 无疑与 *Jonesea* Cocks et Rong(1989)也很相似,它们的主要区别是 *Chonetoidea* 背、腹内壳底的小突起较细长,壳表经常饰有粗、细两组壳纹;而 *Jonesea* 壳内的小突起粗圆,呈乳头状,壳表饰棱形壳线(或棱形壳纹)。

分布及时代:世界各地;中奥陶世晚期(达瑞威尔期晚期)至早志留世(Llandoverian)。

简单似戟贝(新种) *Chonetoidea simplex* Zeng, Zhang et Han(sp. nov.)

图版(pl.)18,图(figs.)1-12;图版(pl.)19,图(figs.)1,2

词源:Simplex(英文),简单的,表示新种背、腹内部构造较简单。

描述:贝体很小,通常壳长2.4～3.3mm,壳宽4.3～5.2mm(表13);轮廓半圆形;侧视平凸型或轻微凹凸型,贝体很薄;铰合线直,等于最大壳宽;主端锐角状或近直角状。腹壳缓凸,仅在喙部前方稍微隆起;腹铰合面低,斜倾型;腹三角孔被小假三角板覆盖。背壳平坦或轻微浅凹;背铰合面很低;背三角孔可能覆有背三角双板。壳表饰粗细较均匀壳纹,通常作2～3次插入式增多。假疹壳。

表13 简单似戟贝(新种)介壳测量(单位:mm)
Table 13 Shell measurements of *Chonetoidea simplex* (sp. nov.) (in mm)

采集号 (Coll. No.)	登记号 (Cat. No.)	腹壳(ventral valve)		背壳(dorsal valve)		备注 (remarks)
		长(length)	宽(width)	长(length)	宽(width)	
DH3	HB493	3.2	5.3			
DH3	HB574	3.3	5.2			副型(paratype)
DH3	HB335	1.9	3			
DH2	HB88	2.7	4.3			
DH2	HB424	2.9	4.5			副型(paratype)
DH2	HB425	3.3	5.2			
DH2	HB441	2.7	4.9			
DH2	HB430	3	4.8			
DH2	HB431			2.7	4.2	正型(holotype)
DH2	HB433			3.4	5.2	
WH2	HB149			2.4	4.3	
DH2	HB426			2	3.2	
DH3	HB470			2.9	4	副型(paratype)

腹内:齿板短,延伸在腹肌痕面两外侧;腹肌痕面明显双叶状,中间被显著的短肌隔分开;腹壳底布满较细长、不规则的小突起。

背内:铰窝小,三角状;内铰窝脊短宽,约呈145°～150°角异向展伸;主突起小,双叶型,呈"八"字形,其两前端分别与内铰窝脊后端互相融合;主突起前方具一椭圆形主突起坑;中隔板薄而高,从主突起坑前方开始伸出直至背壳底中前部;缺失肌台;背壳底具许多不规则、较细长的小突起。

讨论:本新种 *Chonetoidea simples* (sp. nov.)无论从其外形还是从其背、腹内构造特征来看,都与 *Aegiromena* Havliček(1961)很相似,它们唯一的区别是本新种 *Chonetoidea simples* (sp. nov.)的背内缺失肌台(Bema)和仅具有单背中隔板,而 *Aegiromena* Havliček 的背内具有肌台和双背中隔板。另外由于新种 *Chonetoidea simples* (sp. nov.)的背、腹内构造显得较为简单而与 *Chonetoidea* 已有各个种背、腹壳底的小突起均具有一定规律的排列方式可以互相区别。

产地层位:湖北宜昌丁家坡、王家湾;上奥陶统顶部五峰组观音桥段(赫南特阶中部)。

三板月贝属(新属)Genus *Trimena* Zeng, Wang et Peng(gen. nov.)

属型种:Genotype *Trimena wangjiawanensis* Zeng, Wang et Peng(gen. et sp. nov.).

词源:Tri(拉丁文),三,表示新属背内具3根高强隔板的小型埃月贝类腕足。

特征简述:壳形像 *Aegiria* Öpik(1933),但是新属 *Trimena*(gen. nov.)的腹启肌痕面为长椭圆形,并呈宽距离双叶型;背内具有高强的背中隔板和2根侧隔板;放射纹简单、稀疏;上述这些特征是 *Aegiria* 所没有的。

描述:贝体小,轮廓半圆形,侧视平凸型或轻微凹凸型;铰合线直,等于最大壳宽;主端锐角状或近直角状。腹壳凸度平缓,仅在喙部前方稍微隆起;腹铰合面低,斜倾型;腹三角孔可能被假三角板覆盖。背壳平坦或轻微浅凹;背铰合面极低,呈线状;背三角孔被背三角板覆盖。壳表饰简单放射纹,作2~3次插入式增多。假疹壳。

腹内:齿板短,异向展伸在腹肌痕面两后侧;腹肌痕面显著,启肌痕面呈长椭圆形,并为宽距离的双叶型;肌隔短而高,位于两启肌痕之间的后部(图版20,图1;插图20-A),闭肌痕不清晰。

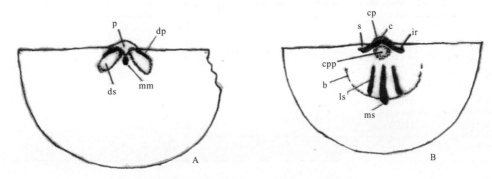

插图20 *Trimena*(gen. nov)的内部构造

Illustr. 20 Interior structures of *Trimena*(gen. nov.)

A-腹内模(据图版20,图1);B-背内模(据图版20,图4a,4b)

A-Ventral internal mold(from pl. 20,fig. 1);B-Dorsal internal mold(from pl. 20,figs. 4a,4b)

b-肌台(bema);c-背三角板(chilidium);cp-双叶状主突起(bilobed cardinal process);cpp-主突起坑(cardinal process pit);dp-齿板(dental plate);ds-启肌痕(diductor scars);ir-内铰窝脊(inner socket ridge);ls-侧隔板(lateral septum);mm-中肌隔(median myophragm);ms-中隔板(median septum);p-假三角板(pseudodeltidium);s-铰窝(socket)

背内:铰窝小,长三角状;内铰窝脊短宽,强烈异向展伸;主突起显著,双叶型,其前端与内铰窝脊后端相融合;主突起前方具一椭圆形主突起坑;中隔板高强,始于主突起坑前方,其前端穿越肌台前缘围脊;两侧隔板高强,位于中隔板两侧边,但其前端不超越肌台前缘围脊(图版20,图4a,4b,插图20-B);肌台椭圆状,围脊隐约可见。

讨论:Zeng(1977)曾鉴定为 *Aegiromena ultima* Marek et Havliček(59页,图版20,图12),以及Rong(1984)曾鉴别为 *Aegiromena convexa* Chang(P. 184,pl. 11,fig. 3),目前对上述2个背内模进行重新辨别,发现它们都具有高强背中隔板和2根侧隔板,这显然不同于 *Aegiromena* Havliček(1961)属征的含义,而应归于 *Trimena*(gen. nov.)的类型。

新属 *Trimena*(gen. nov.)与 *Aegiria* Öpik(1933)或者 *Aegiromena* Havliček(1961)的主要区别是:新属 *Trimena*(gen. nov.)的腹启肌痕面呈长椭圆形,并为宽距离的双叶型,背内具3根高强隔板;而 *Aegiria* 的腹启肌痕面为椭圆形、为近距离的双叶型,背内为单中隔板;*Aegiromena* 的腹启肌痕面也为椭圆形、近距离的双叶型,但背内则为双中隔板。

分布及时代:中国中南部;晚奥陶世末期赫南特期中期(Middle Hirnantian)。

王家湾三板月贝(新属、新种)*Trimena wangjiawanensis* Zeng, Wang et Peng(gen. et sp. nov.)

图版(pl.)19,图(fig.)10;图版(pl.)20,图(figs.)1-4;
插图(Illustr.)20-A,B

1977 *Aegiromena ultima* Marek et Havlíček;Zeng. 59 页,图版 20,图 12。
1984 *Aegiromena convexa* Chang;Rong. P. 148,pl. 11,fig. 3.

词源:Wangjiawan 为化石产地王家湾的汉语拼音;ensis 表示化石来源。

描述:贝体很小,通常壳长 2.9~4mm,壳宽 4.8~6mm(表 14);轮廓半圆形,侧视平凸型或者轻微凹凸型,铰合线直,等于最大壳宽;主端锐角状或近直角状。腹壳凸度低缓,仅在喙部稍微前方轻微隆起;腹铰合面低,斜倾型;腹三角孔可能被假三角板覆盖。背壳平坦或轻微浅凹;背铰合面很低,呈线状;背三角孔被背三角板覆盖。壳表饰稀疏放射纹,作 2~3 次插入式增多。

表 14 王家湾三板月贝(新属、新种)介壳测量(单位:mm)
Table 14 Shell measurements of *Trimena wangjiawanensis*(gen. et sp. nov.)(in mm)

采集号 (Coll. No.)	登记号 (Cat. No.)	腹壳(ventral valve)		背壳(dorsal valve)		备注 (remarks)
		长(length)	宽(width)	长(length)	宽(width)	
WH3	HB371			3	约 4.8	
DH2	HB537	4	6			副型(paratype)
WH3	HB373			2.2	3.6	
WH1	HB370			2.9	4.8	正型(holotype)

背、腹内部构造同属征。

产地层位:湖北宜昌王家湾、丁家坡;上奥陶统顶部五峰组观音桥段(赫南特阶中部)。

直形贝目 Order Orthotetida Waagen,1884
直形贝亚目 Suborder Orthotetidina Waagen,1884
直形贝超科 Superfamily Orthotetoidea Waagen,1884
法顿贝科 Family Fardeniidae Williams,1965
####### 法顿贝亚科 Subfamily Fardeniinae,Williams,1965

法顿贝属 Genus *Fardenia* Lamont,1935

1935 *Fardenia* Lamont. P. 310.
1951 *Fardenia* Lamont;Williams. P. 119.
1965 *Fardenia* Lamont;Williams. H407.
1966 *Fardenia* Lamont;Wang,Jin et Fang. 324 页。
1981 *Fardenia* Lamont;Chang. 563 页。
2000 *Fardenia* Lamont;Williams et Brunton. P. 672.

属型种:Genotype *Fardenia scotica* Lamont,1935.

特征简述:贝体小至中等大,轮廓亚方形,侧视低双凸型;铰合线直,等于最大壳宽;主端近直角状。腹壳微凸;腹铰合面低,斜倾型;腹三角孔被假三角板覆盖,成年壳存在茎孔。背壳微凸;背铰合面低,正倾型;背三角孔被背三角板覆盖。壳表饰粗细不均放射纹,作 2~3 次插入式增加。

腹内:齿板短,作窄角度异向展伸;腹肌痕面不清晰。

背内:铰窝显著,三角形;内铰窝脊粗,轻微相向内弯;主突起短,双叶型;背肌痕面不显著。

比较:*Fardenia* Lamont(1935)与 *Coolinia* Bancroft(1949)很相似,Williams(1965)曾将 *Coolinia* 作为 *Fardenia* 的同义名。这两个属的主要区别是 *Fardenia* 的贝体轮廓呈近方形,主端近直角状,腹壳具茎孔;而 *Coolinia* 的贝体较横宽,轮廓呈半椭圆形,主端为尖角状,腹壳无茎孔,放射纹较稀疏,有时具很弱的同心纹。

分布及时代:中国中南部、欧洲;晚奥陶世赫南特期(Hirnantian)。

苏格兰法顿贝 *Fardenia scotica* Lamont,1935

图版(pl.)20,图(figs.)5,8,9,11

1935 *Fardenia scotica* Lamont. P. 130.
1981 *Fardenia scotica* Lamont;Chang. 653 页,图版 1,图 40。
1983 *Coolinia propinquua*(Meek et worthen,1868);Zeng. 121 页,图版 17,图 12-14。
1984 *Coolinia* sp. Rong. P. 161,pl. 12,figs. 11,14;pl. 14,fig. 9c.
2000 *Fardenia scotica* Lamont;Williams et Brunton;P. 672,figs. 486a-f.

描述:贝体小到中等大,通常壳长 7.5~15.6mm,壳宽 8~18.3mm(表 15);轮廓近方形;侧视低双凸型;铰合线直,等于最大壳宽;主端近直角状。腹壳微凸,仅在喙部附近稍微隆起;腹铰合面低,斜倾型;腹三角孔被隆凸假三角板覆盖(图版 20,图 5)。背壳微凸,仅在喙部附近轻微凸起;背铰合面低,正倾型;背三角孔被背三角板覆盖(图版 20,图 9)。壳表饰放射纹,但粗细不均匀,一级壳纹较粗,二级和三级壳纹依次逐步变细。

表 15 苏格兰法顿贝介壳测量(单位:mm)

Table 15 Shell measurements of *Fardenia scotica* Lamont(in mm)

采集号 (Coll. No.)	登记号 (Cat. No.)	腹壳(ventral valve)		背壳(dorsal valve)	
		长(length)	宽(width)	长(length)	宽(width)
WH2	HB29	7.5	约8		
HH2	IV45705			9	约12.6
WH3	HB294	15.6	18.3		

腹内:齿板短,呈 75°~82°窄角度异向展伸;腹肌痕面不清晰。

背内:铰窝小,三角状;内铰窝脊显著,窄角度异向展伸,并轻微相向内弯(图版 20,图 9);主突起双叶状,两个主突起都短粗;背肌痕面不清晰。

讨论:曾庆銮(1983)和 Rong(1984)都曾经把和当前特征相同的标本分别归于 *Coolinia propinquua*(Meek et Worthen,1868)和 *Coolinia* sp.。目前经过慎重对比,认为改归为 *Fardenia costica* Lamont(1935)更为合适。当前获得该类型的标本虽然也不多,但具有较为完整的腹内膜和背内模标本,从其所显现出来的特征上看,宜昌地区观音桥段(层)确实存在 *Fardenia costica* Lamont(1935),尤其是图版 20,图 5,8,9,11 与 Williams et Brunton(2000),P. 672,fig. 486c,486e 的背、腹内模特征完全可以相比。

产地层位:湖北宜昌黄花场、王家湾;上奥陶统顶部五峰组观音桥段(赫南特阶中部)。

库林贝属 Genus *Coolinia* Bancroft,1949

1949 *Coolinia* Bancroft. P. 7.
1965 *Fardenia* Lamont;Williams. H407.
1983 *Fardenia* Lamont;Zeng. 121 页。

2000　*Coolinia* Bancroft; Williams et Brunton. P. 671.

属型种：Genotype *Orthis？applanata* Salter, 1846.

特征简述：贝体小至中等大，轮廓横半椭圆形，侧视双凸型；铰合线直，等于最大壳宽；主端尖翼状。腹壳凸度低，仅在喙部附近轻微隆起；腹铰合面低，斜倾型；腹三角孔后端被假三角板覆盖，在成年体无茎孔。背壳凸度低缓，仅在喙部附近轻微隆起；背铰合面低，正倾型；背三角孔被大的背三角板覆盖。壳表饰放射纹；同心纹弱，不连续。

腹内：铰齿小；齿板短，异向展伸；腹肌痕面呈扇形。

背内：铰窝小，三角形；内铰窝脊短，异向展伸，而且轻微相向内弯；主突短双叶状，在两主突起之间通常具有中基部结；背肌痕面不显著，但具短中肌隔。

比较：*Coolinia* Bancroft(1949)与 *Fardenia* Lamont(1935)很相似。它们的主要区别是 *Coolinia* 的轮廓呈横半椭圆形，主端呈尖翼状，放射纹较稀疏，粗细较均匀；而 *Fardenia* 的轮廓呈近方形，主端近直角状，放射纹粗细不均匀。

分布及时代：世界各地；晚奥陶世赫南特期(Hirnantian)至志留纪。

库林贝（未定种）*Coolinia* sp.

图版(pl.)20, 图(figs.)6, 7

1983　*Fardenia* sp. Zeng. 121 页, 图版 16, 图 14。

描述：当前仅获得 3 枚标本，3 枚腹内膜（含曾庆銮, 1983, 图版 16, 图 14），1 枚背内模。贝体很小，壳长 1.9～2.1mm，壳宽 2.6～3.8mm；轮廓横半椭圆形；侧视低双凸型；铰合线直，等于最大壳宽；主端锐角状。腹壳凸度低，仅在喙部前方微凸；腹铰合面低，斜倾型；腹三角孔被假三角板覆盖（图版 20, 图 6）。背壳微凸，仅在喙部附近稍微隆起；背铰合面极低，呈线状；背三角孔被背三角板覆盖。放射纹稀疏，但未见同心纹痕迹。

腹内：齿板短，呈 80°角异向展伸；腹肌痕面隐约可见，呈梯形。

背内：铰窝显著，呈三角状；内铰窝脊显著，呈 130°角异向展伸，不相向内弯；主突起短小，呈双球状，并位于铰合线之外；背肌痕面模糊，并且微凹。

讨论：当前标本的壳形、壳饰、腹壳内部构造特征都和 *Coolinia* Bancroft(1949)很接近；但是背内的双叶状主突起位于铰合线之外，内铰窝脊未相向内弯；背肌痕面中间无肌隔等这些特征又与 *Coolinia* 有所不同，因此，将当前的标本暂时处置为 *Coolinia* sp, 待今后获得更多标本后再作进一步鉴别。

产地层位：湖北宜昌王家湾、黄花场；上奥陶统顶部五峰组观音桥段（赫南特阶中部）。

三重贝亚目 Suborder Triplesiidina Moore, 1952

　　三重贝超科 Superfamily Triplesioidea Schuchert, 1913

　　　　三重贝科 Family Triplesiidae Schuchert, 1913

三重贝属 Genus *Triplesia* Hall, 1859

1859　*Triplesia* Hall. P. 44.
1951　*Triplesia* Hall; Williams. P. 104.
1956　*Triplesia* Hall; Cooper. P. 537.
1965　*Triplesia* Hall; Wright. H358.
1966　*Triplesia* Hall; Wang, Jin et Fang. 243 页。
1974　*Triplesia* Hall; Williams. P. 116.
1977　*Triplesia* Hall; Zeng. 48 页。
1980　*Triplesia* Hall; Nikitin. P. 45.
1981　*Triplesia* Hall; Lockley et Williams. P. 55.

1981　*Triplesia* Hall;Chang. 561 页。
1981　*Triplesia* Hall;Zeng. 118 页。
1984　*Triplesia* Hall;Rong. P. 146.
2000　*Triplesia* Hall;Wright. P. 685.

属型种：Genotype *Atrypa extans* Emmons,1842.

特征简述：贝体中等大,轮廓阔卵形;铰合线短,主端阔圆状;前接合缘强烈单褶型;侧视双凸型。腹壳凸度中等,腹中槽发育;腹铰合面窄、高、微弯;腹三角孔小,被假三角板覆盖。背壳凸度中等或者高强;背铰合面窄小,微弯。背中隆显著,两侧区缓凸。壳表光滑,仅有微弱的同心层。

腹内：齿板短,异向展伸;腹三角腔中线具圆柱状肉茎管;腹肌痕面不清晰。

背内：铰窝小,三角状;内铰窝脊极短或者无;腕基突起耸立;主突起极为强壮,主突起茎长,其末端分成"叉"状,并远远超伸于铰合线之外;背肌痕面不清晰。

比较：*Triplesia* Hall(1859)与 *Cliftonia* Foerste(1909)很相似,它们的主要区别是 *Triplesia* 的壳表光滑,同心层微弱;而 *Cliftonia* 的壳表饰放射线和叠瓦状同心层。

分布及时代：中国中南部、北美、西欧;奥陶纪至志留纪兰多维列世。

宜昌三重贝 *Triplesia yichangensis* Zeng

图版(pl.)20,图(fig.)10;图版(pl.)21,图(figs.)1-6

1977　*Triplesia yichangensis* Zeng. 48 页,图版 16,图 1-3。
1981　*Triplesia yichangensis* Zeng;Chang. 561 页,图版 1,图 16。
1983　*Triplesia yichangensis* Zeng;Zeng. 118 页,图版 17,图 15-17。
1984　*Triplesia yichangensis* Zeng;Rong. P. 146, pl. 10, figs. 6, 11.

描述：贝体中等大,通常壳长 12~16mm,壳宽 8.9~13.5mm(表 16);轮廓长卵形;铰合线短,主端阔圆状,最大壳宽位于贝体中前部;前接合缘单褶型;侧视双凸型。腹壳凸度不大,腹铰合面窄而高,微弯;茎孔小(图版 21,图 6);腹中槽发育,始于喙部附近,向前加宽加深,并向前缘突伸呈"舌状",槽底宽平,两侧区壳面高于槽底,凸度平缓。背壳凸度稍微大于腹壳;背铰合面窄小,高而微弯;背中隆显著,始于喙部前方,向前加宽加高,并向前缘突伸;两侧区壳面低于背中隆,凸度缓和。壳表光滑,仅在中前部饰微弱同心层。

表 16　宜昌三重贝介壳测量(单位:mm)
Table 16　Shell measurements of *Triplesia yichangensis* Zeng(in mm)

采集号 (Coll. No.)	登记号 (Cat. No.)	腹壳(ventral valve)		背壳(dorsal valve)	
		长(length)	宽(width)	长(length)	宽(width)
WH1	HB78	16	13.5		
DH2	HB632	14	11.5		
DH2	HB633			15.2	13.9
DH2	HB93			14	13.5
WH2	HB608	约12	8.9		

腹内：齿板短,呈 55°~65°角异向展伸;腹三角腔中间具有圆柱状肉茎管(图版 21,图 2,6);腹肌痕面不清晰。

背内：铰窝显著,呈三角状;腕基突起粗壮,耸立;主突起极为强大,主突起茎长,其末端分叉,叉枝也很长,形成显著的叉状(图版 21,图 5);背肌痕面不清晰。

产地层位：湖北宜昌丁家坡、分乡、王家湾；上奥陶统顶部五峰组观音桥段（赫南特阶中部）。

分乡三重贝 *Triplesia fenxiangensis* Yan

图版(pl.)21,图(figs.)7-12

1978 *Triplesia fenxiangensis* Yan. 219页,图版62,图8,9,24。
1981 *Triplesia fenxiangensis* Yan;Chang. 561页,图版1,图21。
1981 *Triplesia sanxiaensis* Chang. 561页,图版1,图22-24。
1983 *Triplesia fenxiangensis* Yan;Zeng. 图版17,图18,19。

描述：贝体中等大,通常壳长11~14mm,壳宽12~14.5mm（表17）；壳长稍微短于壳宽,轮廓近圆形；侧视双凸型；铰合线短,主端圆弧状；最大壳宽位于贝体中部。腹壳凸度低；腹中槽浅,始于喙部前方,向前逐步加宽；两侧区壳面平缓。背壳凸度稍微大于腹壳；背中隆低,仅在壳面中前部稍微隆起。壳表光滑,仅具有微弱的同心线或同心层。

表17 分乡三重贝介壳测量(单位:mm)
Table 17 Shell measurements of *Triplesia fenxiangensis* Yan(in mm)

采集号 (Coll. No.)	登记号 (Cat. No.)	腹壳(ventral valve)		背壳(dorsal valve)	
		长(length)	宽(width)	长(length)	宽(width)
HH2	HB729	11.5	12		
FH2	HB730			11	12.5
WH2	HB38	3.5	3		
WH3	HB651	14	14.5		

腹内：齿板短,小角度异向展伸；腹三角腔中间具圆柱状肉茎管（图版21,图7）。

背内：腕基突起强,耸立；主突起强大,主突起茎较短,其后端分枝呈叉状,叉枝很长（图版21,图9）。

比较：*Triplesia fenxiangensis* Yan(1978)与*Triplesia yichangensis* Zeng(1977)的主要区别是*Triplesia fenxiangensis* 的壳长短于壳宽,贝体轮廓近圆形；腹中槽和背中隆相对较短；而*Triplesia yichangensis* 的壳长大于壳宽,轮廓呈长卵形,腹中槽和背中隆较长,较显著,背内主突起茎较长。

产地层位：湖北宜昌黄花场、分乡、王家湾；上奥陶统顶部五峰组观音桥段（赫南特阶中部）。

克利夫通贝属 Genus *Cliftonia* Foerste,1909

1909 *Cliftonia* Foerste. P. 81.
1951 *Cliftonia* Foerste;Williams. P. 105.
1963 *Cliftonia* Foerste;Wright. P. 761.
1965 *Cliftonia* Foerste;Wright. H358.
1967 *Cliftonia* Foerste;Marek et Havliček. P. 281.
1968 *Cliftonia* Foerste;Bergström. P. 11.
1977 *Triplesia* Hall;Mitchell. P. 66.
1980 *Cliftonia* Foerste;Nikitin. P. 46.
1981 *Cliftonia* Foerste;Chang. 561页。
1983 *Cliftonia* Foerste;Zeng. 119页。
1984 *Cliftonia* Foerste;Rong. P. 146.
2000 *Cliftonia* Foerste;Wright. P. 685.

属型种：Genotype *Cliftonia striata* Foerste(1909).

修订后的特征简述：由于 Foerste(1909)建立 *Cliftonia* 属时描述过于简单，尤其是背内的许多构造不明确。当前获得较清晰的背内模标本，现将 *Cliftonia* 的属征补充描述如下：

贝体中等大，轮廓亚圆形至长卵形，侧视不等双凸型；铰合线短，最大壳宽位于贝体中前部；主端阔圆；前接合缘单褶型。腹壳凸度较低；腹喙部微弯；腹铰合面窄而高，斜倾型；腹中槽强、弱不等。背壳凸度大于腹壳；背壳喙部微弯；背铰合面窄而高，正倾型；背中隆强、弱不等。壳表饰稀疏、较粗的放射线，并被叠瓦状同心层所穿越。

腹内：齿板短、薄板状，呈小角度异向展伸；腹三角腔中间具一圆柱状肉茎管；腹肌痕面不清晰。

背内：铰窝小，短三角状；内铰窝脊极短或者无；腕基突起粗壮，耸立；主突起极强大，主突起茎短粗，其后端分叉，叉枝很长，使主突起呈叉状；在老年介壳内，腕基支板显著，近平行向前延伸；在腕基支板之间的前部具一圆形深坑（可能为主突起坑）；在该圆形深坑两前侧各具一条深沟（后侧沟），并呈 103°角左右异向展伸（图版 56，图 1，2；插图 21）；在腕基支板两外侧各具一块近方圆形、轻微隆起的光滑壳底（可能为后对闭肌痕面）；在两条深沟之前具一伞形平台（可能为前闭肌痕），其前部具 2 根纵侧脊；在 2 根纵侧脊的外侧各具一个椭圆形凹坑。

比较：*Cliftonia* Foerste(1909)与 *Triplesia* Hall (1859)的壳形、腹内构造，以及背内主基形态都非常相似，它们的主要区别是 *Cliftonia* 的壳表饰放射线，并且具叠瓦状同心层，背内构造很复杂（插图 21）；而 *Triplesia* 的壳表光滑，仅在壳表中前部具有同心层，背内未见到图版 22，图 4，7 那样的构造。

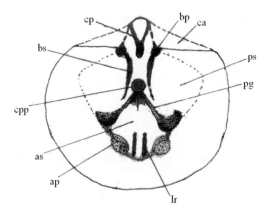

插图 21　*Cliftonia oxoplecioides* Wrigth 的背壳内部构造（据图版 22，图 4，7）

Illustr. 21　Interior structures of dorsal valve of *Cliftonia oxoplecioides* Wrigth(from pl. 22,figs. 4,7)

ap－前坑(anterior pit)；as－前闭肌痕(anterior adductor scars)；bp－腕基突起(brachiophore process)；bs－腕基支板(brachiophore support)；ca－铰合面(cardinal area)；cp－主突起(cardinal process)；cpp－主突起坑？(cardinal process pit?)；lr－侧脊(lateral ridges)；pg－后侧沟(posterior lateral groove)；ps－后闭肌痕(posterior adductor scars)

分布及时代：亚洲东部、北美洲、西欧；晚奥陶世赫南特期至志留纪兰多维列世。

似锐重贝克利夫通贝 *Cliftonia oxoplecioides* Wright

图版(pl.)22,图(figs.)1-12；图版(pl.)56,图(figs.)1,2；
插图(Illustr.)21

1963　*Cliftonia oxoplecioides* Wright. P. 761,pl. 109,figs. 3,4,6,8,10,11.
1967　*Cliftonia oxoplecioides* Wright；Marek et Havlíček. P. 281,pl. 11,figs. 18,21.
1981　*Cliftonia obovata* Chang. 561 页，图版 1，图 25,26。
1983　*Cliftonia oxoplecioides* Wright；Zeng. 119 页，图版 17，图 34。
1984　*Cliftonia obovata* Chang；Rong. P. 147, pl. 10,figs. 3-5,7,8,10,12.
2006　*Cliftonia oxoplecioides* Wright. Rong. 294 页，图版 2，图 14,16。

描述：贝体中等大，通常壳长 10.5～17mm，壳宽 12～19mm（表 18），而且壳宽总是大于壳长；轮廓亚圆形；侧视双凸型；铰合线短于最大壳宽，最大壳宽位于贝体中部；主端圆弧状；前接合缘单褶型。腹壳凸度较低；腹铰合面窄、高，微弯，斜倾型；腹中槽始于喙部前方，向前逐步加宽、加深，延伸至前缘。背壳凸度稍微大于腹壳；背铰合面窄、高，正倾型；背中隆始于喙部前方，向前逐步加宽、加高，延伸至前缘。壳表饰较粗的放射线，并被叠瓦状同心层所贯穿，但有时放射线或同心层较弱。

腹内：齿板短，薄板状，呈 60°～70°角度异向展伸；腹三角腔中间具一圆柱状肉茎管（图版 22，图 2）；

腹肌痕面不清晰。

背内：铰窝小，短三角状；内铰窝脊极短，或者无；腕基突起粗壮，耸立；主突起强大，茎部短粗，其后端分叉呈叉状，叉枝长，向后方突伸（图版22，图7，9，11，12）。腕基支板和背壳底构造形态的发育程度多变，通常在青年贝体中都不发育。但是，在老年贝体中腕基支板很显著，呈近平行向前延伸；在腕基支板前方具一圆形深坑（可能为主突起坑）（图版56，图1；图版22，图4，7）；在该圆形深坑的两前侧各具一条深沟，并呈103°角左右异向展伸（图版56，图1，2；插图21）；在腕基支板两外侧各具一轻微隆起、近方圆形的光滑壳底（可能为后对闭肌痕）；在两深沟之前具一伞形、轻微隆起的平台，其中前部具2根短脊（该平台可能为前对闭肌痕）；在伞形平台的两前侧还各具一方圆形凹坑（图版56，图1；插图21）。

表18 似锐重贝克利夫通贝介壳测量（单位：mm）
Table 18 Shell measurements of *Cliftonia oxoplecioides* Wright (in mm)

采集号 (Coll. No.)	登记号 (Cat. No.)	腹壳(ventral valve)		背壳(dorsal valve)	
		长(length)	宽(width)	长(length)	宽(width)
DH3	HB669	16.3	16.8		
DH3	HB668	16.2	16.4		
DH2	HB117	13	14		
DH2	HB72			13	14.2
WH2	HB167			3.8	5.2
WH3	HB658			15	16
HH1	IV45719	10.5	12		
WH2	HB95			12	13
DH3	HB332			约17	19
HH1	HB726			8.5	9

讨论：常美丽（1981）建立的 *Cliftonia obovata* Chang 与 *Cliftonia oxoplecioides* Wright（1963）没有什么差别，因此前者应为后者的同义名。

产地层位：湖北宜昌丁家坡、王家湾、黄花场；上奥陶统顶部五峰组观音桥段（赫南特阶中部）。

拉长克利夫通贝（新种）*Cliftonia elongata* Zeng, Zhang et Peng (sp. nov.)
图版(pl.)45，图(figs.)1-5

词源：Elongata（英文），拉长，伸长，表示新种的贝体拉长成长卵形。

1983 *Cliftonia obovata* Chang; Zeng. 119页，图版17，图20，21。

描述：贝体中等大，壳长11.5～23mm，壳宽10～22mm（表19）；轮廓长卵形；侧视不等双凸型；铰合线短于最大壳宽；最大壳宽位于贝体中部；主端阔圆；前接合缘单褶型。腹壳凸度较低；腹铰合面窄、高，斜倾型；腹中槽显著，始于喙部稍微前方，向前逐步加大、加深，并向前缘突伸呈舌状，槽底平坦。背壳凸度大于腹壳；背铰合面窄、高，但低于腹壳，正倾型；背中隆显著，始于喙部稍微前方，向前逐步加宽、加高，并向前缘突伸呈山脊状。壳表饰低圆放射线和微弱同心线。

腹内：齿板短，薄板状，约呈55°角异向展伸；腹三角腔中部具一圆柱状肉茎管；腹肌痕面不清晰。

表 19 拉长克利夫通贝(新种)介壳测量(单位:mm)
Table 19 Shell measurements of *Cliftonia elongata* (sp. nov.) (in mm)

采集号 (Coll. No.)	登记号 (Cat. No.)	腹壳(ventral valve)		背壳(dorsal valve)		备注 (remarks)
		长(length)	宽(width)	长(length)	宽(width)	
HK1	IV45714	11.5	10			正型(holotype)
HK1	HB724	12	10.5			
HK1	IV45715			11.5	10	副型(paratype)
WH1	HB671			17	约 13	
WH2	HB672			23	22	

背内:铰窝小,短三角状;内铰窝脊极短,或者无;腕基突起粗壮,耸立;主突起强大,茎部短粗,其后端分叉呈叉状,叉枝向后突伸;背肌痕面不清晰。

讨论:本书作者之一(曾庆銮,1983,119 页,图版 17,图 20,21)的标本曾归为 *Cliftonia obovata* Chang(1981),现经过重新鉴别和比较,它们应另立新种。本新种 *Cliftonia elongata*(sp. nov.)与 *C. oxoplecioides* Wright 的主要区别是 *C. elongata*(sp. nov.)的贝体轮廓拉长成长卵形,壳长总是大于最大壳宽;而 *C. oxoplecioides* Wright 的贝体横宽,轮廓呈亚圆形,壳长总是短于最大壳宽。而常美丽(1981)建立的 *C. obovata* Chang 贝体轮廓为亚圆形,壳长远短于壳宽,与 *C. oxoplecioides* Wright 没有什么区别,应为后者的同义名。

产地层位:湖北宜昌黄花场、王家湾;上奥陶统顶部五峰组观音桥段(赫南特阶中部)。

小嘴贝纲 Class Rhynchonellata Williams et others,1996
 正形贝目 Order Orthida Schuchert et Cooper,1932
  正形贝亚目 Suborder Orthidina Schuchert et Cooper,1932
   褶正形贝超科 Superfamily Plectorthoidea Schuchert et LeVene,1929
    弓正形贝科 Family Toxorthidae Rong,1984

弓正形贝属 Genus *Toxorthis* Temple,1968

1968 *Toxorthis* Temple. P. 20.
1970 *Toxorthis* Temple;Temple. P. 11.
1979 *Toxorthis* Temple;Rong. 1 页。
1983 *Toxorthis* Temple;Zeng. 115 页。
1984 *Toxorthis* Temple;Rong. P. 126.
1987 *Toxorthis* Temple;Temple. P. 29.
2000 *Toxorthis* Temple;Williams et Harper. P. 782.

属型种:Genotype *Toxorthis proteus* Temple(1968).

特征简述:贝体很小,轮廓横椭圆形;侧视低双凸型;铰合线直,等于或稍微短于最大壳宽;主端锐角或钝角状。腹壳凸度低,纵中线稍微隆起,但不足为腹中隆;腹铰合面中等高,斜倾型。背壳缓凸,纵中线轻微凹下,但不足为背中槽;背铰合面低,正倾型。壳表饰稀疏放射线,同心纹弱、稀少。

腹内:铰齿低脊状;齿板沿着腹肌痕面周围相向延伸,并会合形成一个悬空匙形台;无腹中隔板。

背内:主基短,而且很开阔;铰窝显著,窄沟状;内铰窝脊显著,强烈异向展伸,几乎与铰合线平行;无腕基支板;背窗台短小,隆升高于壳底;主突起小,双叶型;背肌痕面不清晰;无背中隔板。

分布及时代:中国中南部、英国;晚奥陶世赫南特期中期至志留纪兰多维列世。

奇异弓正形贝 *Toxorthis mirabilis* Rong

图版(pl.)33,图(figs.)9-11

1979 *Toxorthis mirabilis* Rong.2页,图版1,图4,5。
1983 *Toxorthis mirabilis* Rong;Zeng.115页,图版15,图16-18。
1984 *Toxorthis mirabilis* Rong;Rong.P.127,pl.1,figs.1-13,text-fig.8.

描述:贝体很小,通常壳长1.6~1.7mm,壳宽2.7~2.9mm(表20);轮廓横椭圆形;侧视低双凸型;铰合线直,稍微短于最大壳宽;主端钝角状。腹壳凸度低,仅在纵中线轻微隆起,但不足为腹中隆;背壳凸度平缓,仅在纵中线轻微凹下,但不足为背中槽;壳表饰稀少放射线;同心纹少,而且微弱。

表20 奇异弓正形贝介壳测量(单位:mm)
Table 20 Shell measurements of *Toxorthis mirabilis* Rong(in mm)

采集号 (Coll. No.)	登记号 (Cat. No.)	腹壳(ventral valve)		背壳(dorsal valve)	
		长(length)	宽(width)	长(length)	宽(width)
DH2	HB209	1.7	2.8		
DH2	HB210	1.6	2.7		
FH2	HB750			1.7	2.9

腹内:齿板相向沿着腹肌痕面周围延伸,并会合形成一个悬空的匙形台;无腹中隔板。

背内:主基很开阔,铰窝窄沟状,两者相隔甚远;内铰窝脊显著,强烈异向展伸,几乎与铰合线平行;背窗台浅小,高于背壳底;主突起低、小、双叶型;背肌痕面不清晰;无背中隔板。

比较:*Toxorthis mirabilis* Rong(1979)与 *T. proteus* Temple(1968)的主要区别是:*T. mirabilis* 的贝体轮廓为横椭圆形,主端呈钝角状;而 *T. proteus* 的贝体轮廓更加横宽,主端尖翼状。

产地层位:湖北宜昌丁家坡、分乡、棠垭;上奥陶统顶部五峰组观音桥段(赫南特阶中部)。

德姆贝亚目 Suborder Dalmanellidina Moore,1952
  德姆贝超科 Superfamily Dalmanelloidea Schuchert,1913
    德姆贝科 Family Dalmanellidae Schuchert,1913
      德姆贝亚科 Subfamily Dalmanellinae Schuchert,1913

德姆贝属 Genus *Dalmanella* Hall et Clarke,1892

1892 *Dalmanella* Hall et Clarke.P.205.
1955 *Dalmanella* Hall et Clarke;Cooper.P.948.
1963 *Dalmanella* Hall et Clarke;Williams.P.380.
1965 *Dalmanella* Hall et Clarke;Williams et Wright.H333.
1965 *Dalmanella* Hall et Clarke;Temple.P.383.
1966 *Dalmanella* Hall et Clarke;Wang,Jin et Fang.186页。
1967 *Dalmanella* Hall et Clarke;Marek et Havlíček.P.280.
1968 *Dalmanella* Hall et Clarke;Bergström.P.8.
1974 *Dalmanella* Hall et Clarke;Rong.196页。
1977 *Dalmanella* Hall et Clarke;Havlíček.P.136.
1979 *Dalmanella* Hall et Clarke;Rong.2页。
1980 *Dalmanella* Hall et Clarke;Nikitin.P.38.

1982 *Dalmanella* Hall et Clarke；Fu. 108 页。
1983 *Dalmanella* Hall et Clarke；Zeng. 117 页。
1984 *Dalmanella* Hall et Clarke；Rong. P. 128.
2000 *Dalmanella* Hall et Clarke；Harper. P. 783.
2006 *Dalmanella* Hall et Clarke；Rong. 2 页。

属型种：Genotype *Orthis testudinaria* Dalman，1828.

特征简述：贝体小，轮廓亚圆形，侧视近平凸型或低腹双凸型；铰合线直，短于最大壳宽；主端圆弧状；前接合缘近直缘型或轻微、开阔单槽型。腹壳凸度较强，纵中部呈低脊状；腹铰合面微弯，斜倾型；腹三角孔洞开。背壳凸度低缓或近于平凸，纵中部微凹，呈很浅而开阔的背中槽；背铰合面低，正倾型；背三角孔洞开。壳表饰密型放射线。

腹内：铰齿小，锥状；齿板发育，延伸在腹肌痕两侧；在铰齿内侧各具一个深的腕基窝；腹肌痕面亚三角形；闭肌痕直长，居中；启肌痕较长，月牙形，位于闭肌痕两外侧。

背内：背三角腔浅，主基密集。铰窝小，短三角形；腕基短小，耸立，其后边缘与铰合线平行；腕基支板发育，近平行向前延伸，有时其前端轻微相向汇集在粗宽背中隔脊两后侧缘上；主突起显著，单脊状，茎部较长，冠部有时具小齿状突起；背中隔脊强弱不一；背肌痕面发育程度多变。

比较：*Dalmanella* 与 *Paucicrura* Cooper(1956)最为相似，它们的主要区别是：*Dalmanella* 的贝体轮廓为亚圆形，壳表饰密型放射线，腹肌痕面亚三角形，主突起单脊状，冠部有时具小齿状突起；而 *Paucicrura* 的贝体轮廓呈亚方形，壳表饰簇型放射纹，腹肌痕面双叶状，主突起呈三叶型。

分布及时代：世界各地；中奥陶世大坪期至志留纪兰多维列世。

龟形德姆贝 *Dalmanella testudinaria* (**Dalman**)

图版(pl.)23，图(figs.)1-11；图版(pl.)24，图(figs.)1-12

1828 *Orthis testudinaria* Dalman. P. 115, pl. 2, figs. 4a-e.
1965 *Dalmanella testudinaria* (Dalman)；Williams et Wright. H333, fig. 212, 3a-3e.
1965 *Dalmanella testudinaria* (Dalman)；Temple. P. 383, pl. 3, figs. 1-7；pl. 4, figs. 1-6；pl. 5, figs. 1-7；pl. 6, figs. 1-7.
1967 *Dalmanella testudinaria* (Dalman)；Marek et Havliček. P. 280, pl. 2, figs. 1-4.
1968 *Dalmanella testudinaria* (Dalman)；Bergström. P. 8, pl. 2, fig. 5.
1974 *Dalmanella testudinaria* (Dalman)；Rong. 196 页，图版 92，图 22-24。
1977 *Dalmanella testudinaria* (Dalman)；Havliček. P. 137, pl. 32, figs. 1-15, 18, 19, 23；pl. 56, fig. 13.
1979 *Dalmanella testudinaria* (Dalman)；Rong. 3 页，图版 1，图 3，16。
1980 *Dalmanella testudinaria* (Dalman)；Nikitin. P. 38, pl. 12, figs. 1-17.
1982 *Dalmanella testudinaria* (Dalman)；Fu. 109 页，图版 32，图 22-23。
1983 *Dalmanella testudinaria* (Dalman)；Zeng. 117 页，图版 15，图 14，15。
2000 *Dalmanella testudinaria* (Dalman)；Harper. P. 783, fig. 566, 2a-2f.
2006 *Dalmanella testudinaria* (Dalman)；Rong. 294 页，图版 2，图 13。

描述：贝体小，通常壳长 4～5.8mm，壳宽 4.8～6.7mm(表 21)；轮廓亚圆形；侧视近于平凸型或者低腹双凸型；铰合线直，短于最大壳宽，最大壳宽位于贝体中部；主端阔圆状；前接合缘近于直缘型或者极浅、开阔的单槽型。腹壳凸度稍微强些，纵中部隆起较大，呈低脊状；腹喙微弯；腹铰合面中等发育，斜倾型；腹三角孔洞开。背壳凸度低，平坦，纵中部微凹，有时形成开阔、浅背中槽；背铰合面低于腹铰合面，正倾型；背三角孔洞开。壳表饰圆线状放射线，通常作 2～3 次分叉。疹壳，而且壳壁上的盲孔很粗、密，并且沿着壳线间隙沟排列成行(图版 24，图 5-7)。

腹内：铰齿小，锥状，耸立；齿板中等发育，沿着腹肌痕两侧延伸；腹肌痕面清晰可见，为较典型的亚三角形类型(图版 23，图 4)；闭肌痕居中，呈梯形状，其前缘较直，并且稍微长于其两侧的启肌痕；启肌痕较大，呈月牙状(图版 23，图 1)。

表 21 龟形德姆贝介壳测量(单位:mm)
Table 21　Shell measurements of *Dalmanella testudinaria* (Dalman) (in mm)

采集号 (Coll. No.)	登记号 (Cat. No.)	腹壳(ventral valve)		背壳(dorsal valve)	
		长(length)	宽(width)	长(length)	宽(width)
DH2	HB409	5	6.2		
WH2	HB324	5.2	6.2		
WH2	HB234	4.1	5.2		
DH2	HB148	6.8	8.9		
DH2	HB408	4	4.8		
WH3	HB241			3.2	3.7
DH2	HB339			6.9	7.5
WH2	HB246			3.9	4.5
WH1	HB418			3.3	3.9
WH2	HB244			5.3	5.6
WH2	HB245			4.9	5.2
WH2	HB237	5	5.6		
WH1	HB230	5.5	6.8		
WH2	HB243	4.9	5.7		
WH3	HB240	4.3	5		
WH1	HB191			3.8	4.2
WH2	HB236			3.4	4.3
WH2	HB235			3.7	3.9
DH2	HB400			4.3	4.7
DH2	HB239			3.9	4.1
DH2	HB403			5	5.5
WH2	HB256			4.5	5
WH3	HB120			5.8	6.7

背内:主基密集。铰窝小,三角状;腕基极短,耸立;腕基支板长,近平行向前延伸,有时其前端还轻微内弯,并会合在低宽背中隔脊的两后缘(图版 24,图 11,12);主突起单脊状,茎部较长,有时其前端变成线状板向前延伸至与腕基支板等长(图版 23,图 11;图版 24,图 6,9,11);在背三角腔两后侧边的内模各具一个突出小椭圆形铸模,很可能是腹内铰齿内侧的腕基窝在背内模上留下来的铸型(图版 23,图 7a);背肌痕面和背中隔脊都较微弱。

产地层位:湖北宜昌黄花场、丁家坡、王家湾;上奥陶统顶部五峰组观音桥段(赫南特阶中部)。

安尼贝属 Genus *Onniella* Bancroft, 1928

1928　*Onniella* Bancroft. P. 55.
1953　*Onniella* Bancroft; Lindström. P. 132.
1956　*Onniella* Bancroft; Cooper. P. 953.
1963　*Onniella* Bancroft; Williams. P. 404.
1965　*Onniella* Bancroft; Williams et Wright. H336.

1966　*Onniella* Bancroft；Wang,Jin et Fang. 197 页。
1977　*Onniella* Bancroft；Havlíček. P. 124.
1983　*Onniella* Bancroft；Zeng. 118 页。
1984　*Onniella* Bancroft；Rong. P. 129.
2000　*Onniella* Bancroft；Harper. P. 785.

属型种：Genotype *Onniella broeggeri* Bancroft,1928.

特征简述：贝体小至中等大,轮廓亚方形或亚圆形；侧视近平凸型；铰合线直,微短于最大壳宽；主端钝角状或近圆弧状；前接合缘轻微单槽型。腹壳缓凸,顶区稍微隆起；腹铰合面中等高,斜倾型；腹三角孔洞开。背壳近平凸,纵中线微凹,往往形成浅宽背中槽；背铰合面低,正倾型；背三角孔洞开。壳表饰密型放射纹,同心纹微弱。疹壳。

腹内：铰齿小,脊状；齿板短,延伸在腹肌痕面两后侧；腹肌痕面宽亚卵圆形,闭肌痕大,居中,其前端长于两侧的启肌痕,但有时不易区分。

背内：主基强壮,密集。铰窝小,椭圆形,轻微向后斜展,在铰窝底后端经常各具一个小突起；腕基短粗、耸立,轻微斜伸在背三角孔两侧；腕基支板极短或者无；主突起单棱形状,其冠部常常具有数片横小锯齿状突起；背肌痕面呈长椭圆形,前对闭肌痕面较大；背中隔脊低宽,限在背肌痕面内。

比较：*Onniella* 与 *Paucicrura* Cooper(1956)较相似,但 *Onniella* 的壳表饰密型放射纹,腹肌面呈宽卵圆形,背内主突起为单脊状或轻微双叶型；而 *Paucicrura* 的壳饰为簇型放射纹,腹肌痕面呈双叶状,背内主突起为三叶型。

分布及时代：世界各地；中奥陶世大坪期至志留纪兰多维列世。

宜昌安尼贝 *Onniella yichangensis* Zeng

图版(pl.)25,图(figs.)1-12；图版(pl.)26,图(figs.)1-11

1983　*Onniella*? *yichangensis* Zeng(见汪啸风等,1983),118 页,图版 14,图 25,26。
1984　*Onniella*? *yichangensis* Zeng；Rong. P. 129,pl. 2,figs. 1,2,4-9,11；3 和 10.

描述：贝体小,通常壳长 4.3~6.2mm,壳宽 5.6~7.7mm(表 22)；轮廓亚圆形,侧视近平凸型；铰合线直,短于最大壳宽,最大壳宽位于贝体横中部；主端钝圆；前接合缘轻微单槽型。腹壳凸度低,仅在顶区稍微隆起；腹铰合面中等高,斜倾型；腹三角孔洞开。背壳凸度平缓,纵中线轻微凹下,形成浅、宽背中槽；背铰合面较发育,但低于腹铰合面,正倾型；背三角孔洞开。壳表饰细密放射线,一般作 2 次分枝。疹壳。

腹内：铰齿短小,齿板短,异向延伸在腹肌痕面两后侧；腹肌痕面短、宽阔,呈宽卵形,隐约可见闭肌痕较大,居中；启肌痕面相对较短,不包围闭肌痕(图版 25,图 1)。

背内：主基强壮,密集。铰窝小,轻微向背三角孔后侧斜交,有时铰窝底部后端各具一个显著的突起；腕基粗壮、耸立,呈三角状,位于背三角腔两侧边,而且无腕基支板(图版 25,图 2,4),或者极短；背三角腔窄；主突起单棱脊状,或单椭圆脊状,但有时为轻微双叶型(图版 26,图 6-8,10)；背肌痕面呈长椭圆形,并隐约可见前对闭肌痕较大(图版 25,图 2,4,8,12)；背中隔脊低、宽。

讨论：曾庆銮(1983)用一个腹内膜和一个背内模疑问建立 *Onniella*? *yichangensis* 这个种。目前在同一个地区、同一个层位中采集到大量相同的标本,经仔细与 Harper(2000)P. 788,fig. 569,1a-1g 的图影对比,尤其与 fig. 569,1e-1f(*Onniella reuschi*)的主基形态以及其铰窝底后端各具一个突起的情况雷同,因此认定当前这些标本应无疑归入 *Onniella* 这个属。所以,当前将该"?"去掉。另外,Cocks et Fortey(1997),对泰国赫南特动物群研究时,其中 P. 122,pl. 2,figs. 1-7 曾被鉴别为 *Onniella*? *yichangensis* Zeng(1983),这应有存疑,从它们具有较长、小角度异向展伸的腕基支板来看,很可能为 *Drabovia* Havlíček(1950)。

表22　宜昌安尼贝介壳测量(单位：mm)

Table 22　Shell measurements of *Onniella yichangensis* Zeng (in mm)

采集号 (Coll. No.)	登记号 (Cat. No.)	腹壳(ventral valve)		背壳(dorsal valve)	
		长(length)	宽(width)	长(length)	宽(width)
WH1	HB419			4.3	5.6
HH1	HB733			6.2	7.7
WH3	HB613-1			8.2	10
DH2	HB207			7	约8.2
DH2	HB203			5.4	6.7
WH3	HB613-2			5	5.9
DH2	HB202			7	约8.2
DH3	HB513			6	7
WH2	HB255			3.8	4.3
WH2	HB147			5.6	6.8
WH3	HB252			4.8	6.1
DH2	HB206			5.9	6.8
DH2	HB399			4.5	5.2
DH2	HB511			6	6.8
DH2	HB405			3	4
DH2	HB336	5.7	6.3		
DH3	HB315			5	6.2
DH3	HB513			6	7
DH3	HB334			6.2	7.6
DH3	HB524			5.2	6.5

产地层位：湖北宜昌黄花场、丁家坡、王家湾；上奥陶统顶部五峰组观音桥段(赫南特阶中部)。

特鲁西贝属 Genus *Trucizetina* Havlíček, 1974

1974　*Trucizetina* Havlíček. P. 169.
1977　*Trucizetina* Havlíček；Havlíček. P. 138.
1983　*Trucizetina* Havlíček；Zeng. 118 页。
1984　*Trucizetina* Havlíček；Rong. P. 132.
2000　*Trucizetina* Havlíček；Harper. P. 786.

属型种：Genotype *Trucizetina subrotundata* Havlíček, 1974.

特征简述：贝体小，德姆贝形；侧视腹双凸型或近平凸型；背壳具浅宽背中槽。壳表饰簇型或密型放射纹。疹壳。

腹内：铰齿小，椭圆脊状；齿板中等发育，延伸在腹肌痕面两侧；腹肌痕面显著，呈亚三角形；闭肌痕居中，梯形状，稍微短于两侧的启肌痕；启肌痕月牙状。

背内：铰窝小，轻微向背三角腔后侧边斜交；腕基短粗，与背三角腔后侧边斜交；腕基支板发育，异向展伸；主突起窄棱脊状，有时主突起冠部呈微弱双叶型；背肌痕面方圆形或近长方形；背中隔脊低宽。

比较：*Trucizetina*与*Onniella* Bancroft(1928)的外形很相似，它们的主要区别是：*Trucizetina*具有显著、异向展伸的腕基支板，而*Onniella*缺失腕基支板。

分布及时代:欧洲波希米亚和中国中南部;晚奥陶世赫南特期。

宜昌特鲁西贝 *Trucizetina yichangensis* (Zeng)

图版(pl.)27,图(figs.)1-12;图版(pl.)28,图(figs.)1-11

1983 *Drabovia yichangensis* Zeng.115页,图版14,图22。
1983 *Trucizetina subrotundata* Havlíček;Zeng.118页,图版14,图18-21。
1984 *Trucizetina yichangensis*(Zeng);Rong. P.132,pl.2,fig.12.

描述:贝体小,一般壳长3.2~5.8mm,壳宽4~6.5mm(表23),轮廓近圆形,侧视平凸型;铰合线直,短于最大壳宽,最大壳宽位于贝体横中部;主端和侧缘都呈圆弧状;前接合缘轻微单槽型。腹壳缓凸,仅在顶区稍微隆起;腹铰合面中等高,斜倾型。背壳表面平坦,但在纵中线轻微浅凹,形成浅、宽背中槽;背铰合面较低,正倾型。壳表饰放射纹或宽松簇状放射纹。疹壳。

表23 宜昌特鲁西贝介壳测量(单位:mm)
Table 23 Shell measurements of *Trucizetina yichangensis* (Zeng) (in mm)

采集号 (Coll. No.)	登记号 (Cat. No.)	腹壳(ventral valve)		背壳(dorsal valve)	
		长(length)	宽(width)	长(length)	宽(width)
DH3	HB499	7.6	7.9		
DH2	HB356	3.8	4.1		
HH3	HB754			4.6	5.6
WH1	HB249			3.2	4
WH1	HB248			4.3	4.6
DH3	HB508			5.1	6
DH3	HB498			5.2	6.1
WH2	HB619			3.4	4.2
WH2	HB254			4.7	5
DH3	HB549			5.8	6.5
WH2	HB629			3.8	4.8
WH2	HB631			5.6	7
DH3	HB627	3.2	4		
DH2	HB407	5.4	6.5		
DH3	HB115			5.8	7.2
DH3	HB141			4.2	5.4
DH3	HB626			3.4	4.1
DH3	HB151			5.3	6.2
WH2	HB147			5.6	5.8
WH2	HB617			3.6	4.1
DH2	HB550			3	3.8
DH2	HB555			4.2	5.3
WH2	HB618			4.3	5.8

腹内：铰齿小，短、低三角锥状；齿板发育，延伸在腹肌痕面两外侧；腹肌痕面亚三角状；闭肌痕居中，长椭圆形状，其前端稍微短于两侧的启肌痕；启肌痕位于两侧，呈月牙状；但有时腹肌痕面较为模糊。

背内：铰窝小，轻微向背三角腔两后侧边斜交；腕基短粗，也与背三角腔两后侧边斜交（图版27，图3，5-8，10；图版28，图3，4-10）；腕基支板显著，但相对较短、约呈43°~48°小角度异向展伸；主突起在少年期为单脊、刃状（图版27，图4，11；图版28，图5，8），成年期的主突起强壮，椭圆形，冠部作双叶型（图版27，图12；图版28，图10a，10b，11）；背肌痕面在少年期模糊，但在成年期清晰可见，近长方形，并分成前、后两对，而且前对闭肌痕较大，中间被低、宽背中隔脊隔开（图版28，图3，4，6，7）。

讨论：戎嘉余（1984）将曾庆銮（1983）建立的 Drabovia yichangensis 改归为 Trucizetina yichangensis(Zeng)，同时把曾庆銮（1983）识别的 Trucizetina subrotundata Havlíček 也归并于 Trucizetina yichangensis(Zeng)，从当前获得大量的标本证明该方案是正确的。Trucizetina yichangensis 与 T. subrotundata 的主要区别是：T. yichangensis 的腕基支板相对较短，异向展伸的角度相对较小，背肌痕面近长方形等这些特征与 T. subrotundata 的腕基支板较长、异向展伸角度较大、背肌痕近方圆形等特征明显不同。

产地层位：湖北宜昌丁家坡、王家湾；上奥陶统顶部五峰组观音桥段（赫南特阶中部）。

平行特鲁西贝？（新种）*Trucizetina*？ *parallela* Zeng et Zhang(sp. nov.)

图版(pl.)29，图(figs.)1-5

词源：Parallelus(拉丁文)，平行的，表示本新种的腕基支板近平行向前延伸。

描述：贝体小，通常壳长3.7~6.3mm，壳宽4.4~8.2mm（表24）；轮廓亚圆形，侧视近平凸型；铰合线直，短于最大壳宽，最大壳宽位于贝体横中部；主端钝角状或近圆弧状；前接合缘轻微单槽型。腹壳缓凸，仅在顶区轻微隆起；腹铰合面适度高，斜倾型。背壳凸度平坦，纵中线轻微凹下，形成浅、宽背中槽；背铰合面稍微低于腹铰合面，正倾型。壳表饰不很紧密的放射纹，通常作2次分枝。疹壳。

表24 平行特鲁西贝？介壳测量（单位：mm）
Table 24 Shell measurements of *Trucizetina*？ *parallela*(sp. nov.)(in mm)

采集号 (Coll. No.)	登记号 (Cat. No.)	腹壳(ventral valve)		背壳(dorsal valve)		备注 (remarks)
		长(length)	宽(width)	长(length)	宽(width)	
DH3	HB616	6.3	8.2			
DH2	HB512			4	5	正型(holotype)
DH2	HB337			5.2	6.2	
WH2	HB395			2.5	3	
WH2	HB609			3.7	4.4	副型(paratype)

腹内：铰齿短小；齿板延伸在腹肌痕面两外侧；腹肌痕面宽阔，亚三角形，隐约可见闭肌痕居中，前端与其两侧的启肌痕近等长；启肌痕呈月牙状。

背内：主基开阔。铰窝三角形，后端向后斜交于背三角腔两后侧；腕基短、粗，耸立，其后端也斜交于背三角腔两侧边，形成短窄"八"字形；背三角腔为宽阔亚五边形；主突起显著，呈单脊状；腕基支板显著，近平行向前延伸；背肌痕面隐约可见，呈近长方形。

比较：本新种的外形和背内主基形态与 *Trucizetina* Havlíček(1974)的相似，但本新种的腕基支板近平行向前延伸，而 *Trucizetina* 的腕基支板则为异向展伸，因此本新种显然已超出 *Trucizetina* 的含义。本新种的腕基支板特征与 *Dalmanella* Hall et Clarke(1892)的雷同，但本新种的主基开阔，腕基后端与背三角腔两后侧边斜交，这是 *Dalmanella* 所没有的。因此，本新种是具有 *Trucizetina* 与 *Dalmanella* 两个属之间过渡类型的性质。

产地层位：湖北宜昌丁家坡、王家湾；上奥陶统顶部五峰组观音桥段（赫南特阶中部）。

奇异正形贝属 Genus *Mirorthis* Zeng, 1983

1965 *Bancroftina*? Sinclair ; Temple. P. 392.
1977 *Hoderleyella* Bancroft; Havlíček. P. 204.
1983 *Mirorthis* Zeng. 116页（见汪啸风等，1983）。
1984 *Mirorthis* Zeng; Rong. P. 130.
2000 *Mirorthis* Zeng; Harper. P. 785.

属型种：Genotype *Mirorthis mira* Zeng(1983)。

特征简述：贝体小至中等；轮廓亚圆形，侧视低双凸型；铰合线直，短于最大壳宽，最大壳宽位于贝体横中部；主端钝角状或圆弧状；前接合缘直缘型，但为圆弧状。腹壳缓凸，仅在顶区稍微隆起；腹铰合面低，斜倾型；背壳凸度比腹壳更低，接近于平凸；背铰合面比腹铰合面更低，正倾型。壳表饰密型放射纹。疹壳。

腹内：铰齿脊状；齿板显著，相向内弯，延伸在腹肌痕面两外侧；腹肌痕隐约可见，亚卵形，闭肌痕面居中，椭圆形状，其前端与两侧的启肌痕近等长；启肌痕面近月牙形。

背内：主基开阔。铰窝浅、窄长，并轻微向后斜伸（图版31，图6-8，10-11）；腕基发育，其后半段向背三角腔两后侧边斜交，而其前半段则折向背壳底，并向前异向展伸成为外腕基支板（Temple，1965，P.393 和 Rong，1984，P.131 则称为腕基支板）；内腕基支板强（Temple，1965，P.393 和 Rong，1984，P.131 则称为附属支柱，即 ancillary struts），近平行向前延伸，并将极为开阔的背窗腔分隔成3部分；居中的则为通常所称的背窗腔，呈窄、长五边形；而夹于内、外腕基支板之间区域则为附属背窗区，居于背窗腔两外侧，较小，呈近长方形或近长三角形（图版57，图2；插图22）；主突起通常为单脊状，有的呈弱双叶型，有时主突起茎前端还轻微向前延伸呈细脊状；背肌痕面不清楚，背窗腔和附属背窗区可能就是背肌痕面。

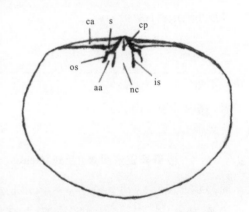

插图22 *Mirorthis mira* Zeng 的背内构造（据图版31，图6；图版57，图2）

Illustr. 22 Showing dorsal interior structures of *Mirorthis mira* Zeng(from pl.31,fig.6;pl.57,fig.2)
aa-附属背窗区(ancillary notothyrial area); ca-铰合面(cardinal area); cp-主突起(cardinal process); is-内腕基支板(inner brachiophore support); nc-背窗腔(nothothyrial cavity); os-外腕基支板(outer brachiophore support); s-铰窝(socket)

讨论：本属曾被疑问置入 *Bancroftina*? Sinclair(1946)(Temple,1965)或被识别为 *Hoderleyella* Bancroft(Havlíček,1977)；但是无论如何，本属背内主基形态很独特，尤其是具有内、外腕基支板，并将极其开阔的背窗腔分隔成3部分，这是 *Bancroftina* 和 *Hoderleyella* 两个属所没有的。

分布及时代：中国中南部、欧洲、非洲摩洛哥；晚奥陶世赫南特期(Hirnantian)。

奇异奇异正形贝 *Mirorthis mira* Zeng

图版(pl.)31,图(figs.)5-11;图版(pl.)57,图(fig.)2;
插图(Illustr.)22

1965 *Bancroftina*? cf. *boučeki*(Havlíček,1950);Temple. P. 392,pl. 7,figs. 2,6.
1983 *Mirorthis mira* Zeng(in Wang et al.). 117 页,图版 14,图 14-17;插图 7。
1983 *Mirorthis yichangensis* Zeng. 117 页,图版 14,图 13。
1984 *Mirorthis mira* Zeng;Rong. P. 131,pl. 4,figs. 1,3,4,6,7,? 9.
2000 *Mirorthis mira* Zeng;Harper. P. 785,fig. 567,2a-2g.

描述:贝体小至中等大,通常壳长 7.3~8.3mm,壳宽 9.3~10.7mm(表 25);轮廓亚圆形,侧视低腹双凸型;铰合线直,短于最大壳宽,最大壳宽位于贝体横中部;主端钝角状或近于圆滑;侧缘圆弧状;前接合缘直缘型,但圆滑。腹壳缓凸,仅在顶区稍微隆起;腹铰合面低,斜倾型;腹三角孔可能洞开。背壳凸度比腹壳更低,壳面接近于平坦,在纵中部轻微凹下,但不足为背中槽;背铰合面低,正倾型。壳表饰细密放射纹,通常作 1~2 次分枝,在近前缘处每 2mm 具 6~7 根。疹壳。

表 25 奇异奇异正形贝介壳测量(单位:mm)
Table 25 Shell measurements of *Mirorthis mira* Zeng(in mm)

采集号 (Coll. No.)	登记号 (Cat. No.)	腹壳(ventral valve)		背壳(dorsal valve)		备注 (remarks)
		长(length)	宽(width)	长(length)	宽(width)	
HH3=HK3	HB740=IV45640	8.3	9.3			
HH3=HK3	HB738=IV45639			7.3	10.7	原来正型 (or. holotype)
DH3	HB144			8.2	?	

背、腹内部构造同属征。

产地层位:湖北宜昌黄花场、丁家坡、王家湾;上奥陶统顶部五峰组观音桥段(赫南特阶中部)。

似奇异正形贝属(新属)Genus *Paramirorthis* Zeng,Wang et Peng(gen. nov.)

属型种:Genotype *Paramirorthis minuta* Zeng,Wang et Peng (gen. et sp. nov.).

词源:Para(英文),相似;*Mirorthis* 为腕足类的一个属名。表示新属与 *Mirorthis* 的关系很密切。

特征简要:贝体很小,轮廓横半圆形,侧视低腹双凸型;壳表饰疏型放射纹。疹壳。齿板短,异向展伸呈"八"字形;腹肌痕面开阔,模糊。主基很开阔、强大、短宽;铰窝细长,向前斜展;腕基长,宽距离异向向前斜伸;腕基支板短,轻微相向内弯或相向斜伸或近平行短距离向前延伸;背三角腔宽圆、短、深;主突起小圆球状或小椭圆形,位于宽圆背三角腔后端。

描述:贝体很小;轮廓横半圆形或横亚长方形;侧视低腹双凸型;铰合线直,近等于最大壳宽;主端近直角状;前接合缘直缘型。腹壳凸度低,仅在腹壳顶部稍微隆起;腹喙短宽;腹铰合面低,斜倾型;腹三角孔可能洞开。背壳凸度低,仅在喙部附近轻微隆起;背铰合面低,正倾型;背三角孔可能洞开。壳表饰简单、近棱脊状、疏型放射纹,通常不分枝或偶作 1 次分枝(图版 57,图 1)。疹壳。

腹内:腹窗腔宽、短;齿板短、直,宽角度异向展伸,呈宽"八"字形;腹肌痕面宽大,模糊。

背内:主基强大、宽、短;铰窝显著,细长沟状,轻微向后斜,并与铰合线轻微斜交(图版 29,图 10;图版 30,图 5-7,8-11);腕基很长,作宽距离异向展伸;腕基支板很短,轻微相向内弯或相向斜伸(图版 30,图 5-11;图版 57,图 1),有的作短距离向前近平行延伸(图版 29,图 8-10);背窗腔短、宽圆、深凹；主突起显著，小圆球状，位于背窗腔后端（图版 29，图 8，10；图版 30，图 6，7，11；图版 57，图 1；插图 23-B），但有时为小单脊状（图版 30，图 8），背肌痕面不清楚，有可能仅限在宽圆的背窗腔内；无背中隔板。

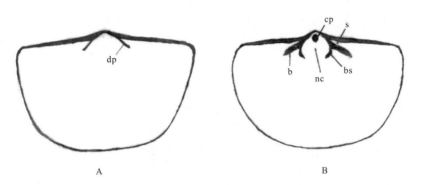

插图 23　*Paramirorthis*（gen. nov.）的内部构造

Illustr. 23　Interior structures of *Paramirorthis*（gen. nov.）

A－腹内模（据图版 30，图 2）；B－背内模（据图版 30，图 6，7；图版 57，图 1）

A－Ventral internal mold（from pl. 30, fig. 2）；B－Dorsal internal mold（from pl. 30, figs. 6,7; pl. 57, fig. 1）

b－腕基（brachiophore）；bs－腕基支板（brachiophore support）；cp－主突起（cardinal process）；dp－齿板（dental plate）；nc－背窗腔（nothothyrial cavity）；s－铰窝（socket）

讨论：新属 *Paramirorthis*（gen. nov.）未建立之前曾经被 Temple（1965）疑问归入 *Bancroftina* ? cf. *boučeki*（Havliček）（pl. 7, fig. 4）或者被 Rong（1984）鉴别为 *Mirorthis mira* Zeng（pl. 3, figs. 2,5,7,9）或者被 Cocks et Fortey（1997）识别为 *Mirorthis mira* Zeng（pl. 2, fig. 8, 但在此应指出的是，该图影应为背内模，而不是腹内模）。当前我们获得许多保存很好的背、腹内模标本。它们的贝体都很小，轮廓为横亚长方形或者横半圆形，壳表饰简单、近棱脊状的放射纹，不分枝或偶然作 1 次分枝；齿板直，强烈异向展伸；主基强大、宽阔；背窗腔短、宽圆、深凹；腕基支板极短，轻微相向内弯，或者相向斜伸；主突起小圆球状，位于背窗腔后端等许多重要特征都与 *Bancroftina* Sinclair（1946）或 *Mirorthis* Zeng（1983）的特征截然不同。证明这些标本不能置入 *Bancroftina* 或 *Mirorthis*，而应另建立 *Paramirorthis*（gen. nov.）这个新属。

分布及时代：亚洲、欧洲；晚奥陶世赫南特期中期。

微小似奇异正形贝（新属、新种）*Paramirorthis minuta* Zeng, Wang et Peng（gen. et sp. nov.）

图版（pl.）29，图（figs.）6-11；图版（pl.）30，图（figs.）1-11；

图版（pl.）31，图（figs.）1-4；图版（pl.）57，图（fig.）1

词源：minute（英文），微小的，表示新种的贝体微小。

描述：贝体微小，通常壳长 1.6～2.5mm，壳宽 2.5～3.8mm（表 26）；轮廓横亚长方形或近横半圆形，侧视低腹双凸型，铰合线直，近等于最大壳宽；主端近直角状；前接合缘直缘型。腹壳凸度低，仅在腹喙至顶区稍微隆起，其他壳面平缓；腹喙宽大，较短而且较低；腹铰合面低，斜倾型；腹三角孔可能洞开。背壳缓凸，仅在背喙附近稍微隆起，其他壳面平缓；背喙宽大，较短，而且较低；背铰合面低，正倾型；背三角孔可能洞开。壳表饰简单、疏型放射纹，一般不分枝，仅偶尔作 1 次分枝。疹壳。

背、腹内部构造特征同属征。

表 26 微小似奇异正形贝介壳测量(单位:mm)
Table 26 Shell measurements of *Paramirorthis minuta* gen. et sp. nov. (in mm)

采集号 (Coll. No.)	登记号 (Cat. No.)	腹壳(ventral valve)		背壳(dorsal valve)		备注 (remarks)
		长(length)	宽(width)	长(length)	宽(width)	
WH2	HB148	1.6	2.5			
WH2	HB187	2.3	3.9			
DH3	HB140			2.3	3.4	
WH1	HB170			2.9	约4	
DH2	HB174	2.1	3.5			副型(paratype)
WH2	HB162	2.6	3.9			
WH2	HB157	2.5	3.8			副型(paratype)
WH2	HB182	2.2	3.8			
WH2	HB146	2.5	3.8			
DH3	HB150			约1.8	约3.1	副型(paratype)
WH2	HB156			1.5	2.5	正型(holotype)
DH2	HB168			约1.9	约2.8	
WH2	HB163			2	3.2	副型(paratype)
WH1	HB171			2.3	3.8	
WH2	HB176			1.4	2.3	
DH2	HB145			1.8		
WH2	HB177			1.8	3.4	
DH2	HB326			2.9	4.2	副型(paratype)
WH1	HB173			1.8	3.1	
WH2	HB165			1.3	2.2	

产地层位:湖北宜昌王家湾、丁家坡;上奥陶统顶部五峰组观音桥段(Hirnantian 中部)。

全形贝超科 Superfamily Enteletoidea Waagen,1884
  德拉勃贝科 Family Draboviidae Havliček,1950
    德拉勃贝亚科 Subfamily Draboviinae Havliček,1950

德拉勃贝属 Genus *Drabovia* Havliček,1950

1950 *Drabovia* Havliček. P. 45(English,P. 116).
1965 *Drabovia* Havliček;Williams et Wright. H330.
1977 *Drabovia* Havliček;Havliček. P. 244.
2000 *Drabovia* Havliček;Harper. P. 826.

属型种:Genotype *Orthis redux* Barrande,1848.

特征简述:贝体小,轮廓亚方形或亚圆形;侧视双凸型;铰合线直,短于最大壳宽;主端钝角状。腹壳缓凸,最大凸度位于顶区;腹铰合面中等高,斜倾型;背壳凸度低,壳面较平坦,纵中部微凹,但不足为背中槽;背铰合面低,正倾型。壳表饰较粗的放射纹。疹壳。

腹内:齿板薄、较长、内弯,延伸在腹肌痕面两外侧;腹肌痕面宽大,亚卵形;在两齿板之间的顶端具有较显著肉茎胼胝。

背内：铰窝小，三角形；背窗腔窄；腕基短、小；腕基支板显著，异向展伸的夹角约为27°~52°；但有的腕基支板轻微内弯；主突起显著，主突起茎较长，冠部小椭圆状；背肌痕面亚方形，分为前后两对，后对闭肌痕的后缘钝尖；背中隔脊显著，限在背肌痕面内；但有的标本背肌痕面和背中隔脊都较为模糊。

比较：*Drabovia* Havlíček(1950)与*Hirnantia* Lamont(1935)的背、腹内部构造特征很相似，它们的主要区别是*Drabovia*的贝体较小，轮廓为亚方形，壳表的放射纹相对较粗、疏，腹肌痕面为宽亚卵形，肉茎胼胝较发育，腕基支板异向展伸的夹角较小，背窗腔较窄，后对闭肌痕的后缘较钝尖；而*Hirnantia*的贝体较大，背壳凸度较强，壳表放射纹较细密，腹肌痕面通常为亚三角形或双叶形，背窗腔较宽，腕基支板异向展伸的夹角较大，背肌痕面相对较短，多为方圆形，后对闭肌痕的后缘呈钝圆状。

分布及时代：中国中南部、欧洲波希米亚；中奥陶世达瑞威尔期至晚奥陶世赫南特期。

丁家坡德拉勃贝(新种)*Drabovia dingjiapoensis* Zeng et Zhang(sp. nov.)

图版(pl.)32,图(figs.)1-10；图版(pl.)33,图(figs.)1-8

词源：Dingjiapo(汉语拼音),丁家坡,为化石产地。

描述：贝体小，通常壳长4~7.5mm，壳宽4.8~6mm(表27)；轮廓方圆形，侧视低双凸型；铰合线直，稍微短于最大壳宽；主端钝角状，有时近于钝圆；前接合缘直缘型。腹壳缓凸，仅在顶区稍微隆起；腹铰合面中等高，斜倾型。背壳凸度低，接近平坦，纵中线中后部微凹，但不足为背中槽；背铰合面低，正倾型。壳表放射纹较粗、疏，作1~2次分枝。疹壳。

表27 丁家坡德拉勃贝介壳测量(单位:mm)

Table 27 Shell measurements of *Drabovia dingjiapoensis* Zeng et Zhang(sp. nov.)(in mm)

采集号 (Coll. No.)	登记号 (Cat. No.)	腹壳(ventral valve)		背壳(dorsal valve)		备注 (remarks)
		长(length)	宽(width)	长(length)	宽(width)	
WH2	HB676	5.6	6.5			
DH2	HB542			7.5	9	
DH3	HB513			6.6	8.2	
WH1	HB231			3	3.4	
WH3	HB601			4.3	5.2	副型(paratype)
WH1	HB229			4.8	5.5	
WH1	HB228			4.8	5.5	正型(holotype)
WH1	HB248			4.3	4.6	
WH2	HB116			4.5	5.2	
DH3	HB551			5	6	
DH2	HB554	6.3	8			
DH3	HB559	4.9	5.4			副型(paratype)
DH3	HB533			4.2	5	
DH2	HB402	4	4.8			
DH3	HB531			9.3	10.2	
WH3	HB695			3.5	4.2	
DH3	HB299			4.5	5.8	
DH3	HB397			6.5	7.6	

腹内：铰齿小，低脊状；齿板薄，中等长，内弯，延伸在腹肌痕面两外侧；腹肌痕面大，宽亚卵形；闭肌痕面居中，长椭圆状，前缘圆弧状，稍微长于其两侧的启肌痕（图版33，图2，4）；一对启肌痕位于闭肌痕两外侧，月牙状；肉茎胼胝较发育。

背内：铰窝小，三角状；腕基短小；腕基支板薄板状，中等长，约呈28°～50°夹角异向向前展伸；主突起呈单细脊状，但有时为微弱双叶型，具有微弱锯齿状小横板（crenulations）（图版32，图5b）；背肌痕面模糊，但有时隐约可见后对闭肌痕面的后缘呈钝尖状（图版32，图8）；背中隔脊不明显，但有的呈低、宽状（图版32，图5a，6－8）。

比较：新种 Drabovia dingjiapoensis(sp. nov.)与属型种 Drabovia redux(Barrande,1848)的主要区别是 D. dingjiapoensis(sp. nov.)的贝体较小（表27），通常的壳宽仅为4.8～6mm，贝体轮廓呈方圆形，背肌痕面不显著；而 D. redux 的贝体较大（通常壳宽为12～18mm），轮廓为横亚方形，背肌痕面和背中脊较发育。

产地层位：湖北宜昌丁家坡、王家湾；上奥陶统顶部五峰组观音桥段（赫南特阶中部）。

德拉勃贝？未定种 Drabovia? sp.

图版(pl.)34，图(fig.)1

描述：当前仅采到一块背内模标本，但它的特征很显著，因此加以描述。贝体小，壳长4.8mm，壳宽5.4mm；轮廓方圆形，凸度低，壳面平坦；铰合线直，短于最大壳宽；主端钝角状；铰合面低，正倾型。壳表饰稀疏放射纹，作1次分枝。疹壳。

背内：背窗腔窄；铰窝底板发育，铰窝很小，在背内模上几乎见不到；腕基很短小；腕基支板发育，相向内弯，但其前端不会合；主突起小单脊状，短；背肌痕面不清晰，背中隔板不显著。

讨论：当前标本的腕基支板相向内弯，这与 Havliček(1950), English, P. 116－118, pl. 10, fig. 8, Drabovia redux 的背内模很相似，几乎雷同，但是目前仅有一块背内模的标本，因此暂时定为 Drabovia? sp.。

产地层位：湖北宜昌丁家坡；上奥陶统顶部五峰组观音桥段（赫南特阶中部）。

小德拉勃贝属 Genus Drabovinella Havliček, 1950

1950　Drabovinella Havliček. P. 50.
1965　Drabovinella Havliček; Williams et Wright. H331.
1977　Drabovinella Havliček; Havliček. P. 240.
1980　Drabovinella Havliček; Havliček et Branisa. P. 36.
1984　Drabovinella Havliček; Rong. P. 136.
2000　Drabovinella Havliček; Harper. P. 826.

属型种：Genotype Orthis draboviensis Barrande, 1879.

特征简要：贝体小至大，轮廓横椭圆形或亚圆形；侧视近等或不等双凸型；前接合缘直缘型；铰合线直，短于最大壳宽；主端钝角状或者钝圆。腹铰合面较高，斜倾型；腹三角孔洞开。背壳凸度较强或者较平坦，在顶区凸度较强；背铰合面适度高，正倾型；背三角孔洞开。壳表饰放射纹。

腹内：齿板薄、长，轻微内弯，延伸在腹肌痕面两外侧；腹肌痕面亚三角形或者窄双叶状。

背内：主基很密集；铰窝和腕基都很小，铰窝底板发育；背窗腔窄；主突起细长，刃状，冠部无锯齿片状（uncrenulated），茎部向前延伸与细弱背中隔板相融合；腕基支板薄、长，其前半部轻微相向内弯，有时还与背中隔板会合；背中隔板与腕基支板等长或者稍微长一些；背肌痕面分成前后两对，前对较大，后对后缘呈尖突状，并包围着腕基支板前部（插图24－A）；但有的背肌痕面不清晰。

比较：Drabovinella 与 Drabovia Havliček(1950) 以及 Hirnantia Lamont(1935)的主要区别是 Drabovinella 的主基很密集；主突起单刃状，其茎部细长，并与背中隔板相融合；腕基支板长，前半部轻

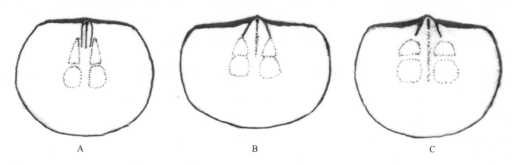

插图 24 背肌痕面对比图(据 Havliček,1977,插图 20)

Illustr. 24 Muscle field comparisons of dorsal valves(from Havliček,1977,fig. 20)

A - *Drabovinella*;B - *Drabovia*;C - *Hirnantia*

微相向内弯;后对闭肌痕后缘尖突,并包围着腕基支板前端;上述 3 个属背内不同特征详见插图 24。

分布及时代:欧洲波希米亚、中国中南部;中奥陶世达瑞威尔期至晚奥陶世赫南特期。

宜昌小德拉勃贝(新种)*Drabovinella yichangensis* Zeng et Zhang(sp. nov.)

图版(pl.)34,图(figs.)2 - 5

1984 *Drabovinella*? sp. nov. Rong. P. 137,pl. 2,figs. 14,15.

词源:Yichang(汉语拼音),宜昌,为化石产地。

描述:贝体小,通常壳长 3.9~5.5mm,壳宽 4~6.2mm(表 28);轮廓亚圆形;侧视低双凸型;铰合线直,短于最大壳宽;最大壳宽位于贝体横中部;主端钝圆。腹壳凸度适度,在顶区凸度稍微强些;腹铰合面中等高,斜倾型。背壳凸度低,接近于平坦,壳面纵中线微凹成浅宽背中槽,但有的不明显;背铰合面适度高,正倾型。壳表饰较粗的放射纹,一般作 1~2 次分枝。疹壳。

表 28 宜昌小德拉勃贝(新种)介壳测量(单位:mm)

Table 28 Shell measurements of *Drabovinella yichangensis*(sp. nov.)(in mm)

采集号 (Coll. No.)	登记号 (Cat. No.)	腹壳(ventral valve)		背壳(dorsal valve)		备注 (remarks)
		长(length)	宽(width)	长(length)	宽(width)	
WH2	HB35	4.2	4.5			
DH2	HB74			约5	5.6	副型(paratype)
WH2	HB53			5.5	6.2	
WH2	HB325			3.9	4	正型(holotype)

腹内:齿板中等长,约呈 45°夹角异向展伸在腹肌痕面的两外侧;腹肌痕面有些模糊,但隐约可见呈窄亚三角形,闭肌痕面直长,居中,稍微长于两侧的启肌痕;启肌痕面较小,呈窄月牙状。

背内:主基密集,铰窝底板发育;铰窝和腕基情况未显露;腕基支板长,后半部近平行向前延伸,而前半部则轻微相向内弯;主突起细长,单刃状,茎部细,向前伸至与腕基支板近等长,或者轻微长一些;背肌痕面不清晰。

比较:新种 *Drabovinella yichangensis*(sp. nov.)的背内特征与 *Drabovinella satrapa* Havliček(1977,P. 242,pl. 12,figs. 14,15;pl. 14,figs. 18 - 21)很相似,它们的主要区别是新种的贝体小,轮廓为

亚圆形，背壳具微弱背中槽，放射纹相对较粗疏；而 D. satrapa 的贝体较大，轮廓为横亚卵圆形，放射纹较细密。

产地层位：湖北宜昌王家湾、丁家坡；上奥陶统顶部五峰组观音桥段（赫南特阶中部）。

赫南特贝属 Genus *Hirnantia* Lamont,1935

1935　*Hirnantia* Lamont. P. 313.
1965　*Hirnantia* Lamont；Williams et Wright. H332.
1965　*Hirnantia* Lamont；Temple. P. 394.
1967　*Hirnantia* Lamont；Marek et Havliček. P. 279.
1968　*Hirnantia* Lamont；Bergström. P. 10.
1974　*Hirnantia* Lamont；Rong. 196 页。
1977　*Hirnantia* Lamont；Havliček. P. 258.
1977　*Hirnantia* Lamont；Zeng. 41 页。
1978　*Hirnantia* Lamont；Yan. 215 页。
1979　*Hirnantia* Lamont；Rong. 5 页。
1980　*Hirnantia* Lamont；Nikitin. P. 43.
1981　*Hirnantia* Lamont；Chang. 559 页。
1982　*Hirnantia* Lamont；Fu. 106 页。
1983　*Hirnantia* Lamont；Zeng. 116 页。
1984　*Hirnantia* Lamont；Rong. P. 137.
1997　*Hirnantia* Lamont；Cocks et Fortey. P. 124.
2000　*Hirnantia* Lamont；Harper. P. 826.
2006　*Hirnantia* Lamont；Rong. 297 页。

属型种：Genotype *Orthis sagittifera* M'Coy,1851.

特征简要：贝体中等至大；轮廓亚圆形或亚椭圆形；侧视近等或背双凸型；前接合缘直缘型或轻微单槽型；铰合线直，短于最大壳宽；主端钝圆状。腹壳凸度中等，最大凸度位于顶区；腹铰合面中等高，斜倾型；腹三角孔洞开。背壳凸度适度，最大凸度位于顶区；背铰合面显著，但低于腹铰合面，正倾型；背三角孔洞开。壳表饰密型放射纹，纹线顶圆滑；同心纹微弱。疹壳。

腹内：齿板短、粗，异向展伸在腹肌痕面两后侧；腹肌痕面显著，腹肌痕围脊强，但肌痕形态多变，多为亚三角形，少数为亚卵形或亚心脏形。

背内：铰窝显著，呈短三角状；腕基短、粗，耸立；腕基支板短，多为70°夹角异向展伸；背窗台短、低，多呈三角状；主突起短、小，冠部有时具小锯片状的横小板，茎部有时向前延伸成细脊状；中肌隔和横肌隔发育，并将背肌痕面分成前后两对；背肌痕面显著，呈方圆形，前后两对闭肌痕的大小多变，但有的背肌痕不清晰。

讨论：由于 *Hirnantia* Lamont(1935)建立的比较早，特征不明确，它的属型种 *Hirnantia sagittifera* (M'Coy)先后曾被多位研究者描述，造成其含义过广，因此，Harper(2000)将以往的 *H. sagittifera* (M'Coy)分为3种不同的类型。本书为便于区分和描述，将该3种类型分别称为：*Hirnantia sagittifera* (M'Coy)morph. Scotland Harper(1989)，*Hirnantia sagittifera* (M'Coy)morph. Bohemia Havliček (1977)和 *Hirnantia sagittifera* (M'Coy)morph. Poland Temple(1965)3个形态种，其中后2个形态种在宜昌地区也有大量标本被发现。

分布及时代：世界各地；晚奥陶世至志留纪兰多维列世。

箭形赫南特贝波希米亚形态种 *Hirnantia sagittifera* (M'Coy,1851)
morph. Bohemia Havliček(1977)

图版(pl.)35,图(figs.)1-12;图版(pl.)36,图(figs.)1,2;
插图(Illustr.)25

1977 *Hirnantia sagittifera* (M'Coy);Havliček.P.266,pl.39,figs.16,17,19,21-23.
2000 *Hirnantia sagittifera* (M'Coy);Harper.P.826,fig.603,1d,1e.

描述:贝体大,通常壳长 24～28mm,壳宽 31～34mm(表 29);轮廓亚圆形;侧视近等双凸型;铰合线直,短于最大壳宽,最大壳宽位于贝体横中部;主端钝圆;前接合缘直线型。腹壳凸度中等,最大凸度位于顶区;腹铰合面适度高,斜倾型。背壳凸度中等,顶区凸度较强,整个壳面凸度较均匀;背铰合面适度高,但低于腹铰合面,正倾型。壳表饰细密放射纹,纹顶圆滑,通常作 2 次分枝。同心纹稀少,微弱。疹壳。

表 29 箭形赫南特贝波希米亚形态种介壳测量(单位:mm)
Table 29 Shell measurements of *Hirnantia sagittifera* (M'Coy) morph. Bohemia(in mm)

采集号 (Coll. No.)	登记号 (Cat. No.)	腹壳(ventral valve)		背壳(dorsal valve)	
		长(length)	宽(width)	长(length)	宽(width)
WH3	HB126	24	32		
WH1	HB81		约31		
WH2	HB66	26	31		
WH3	HB51	27	32		
WH3	HB9			19	25
WH3	HB131			28	34
WH2	HB4	20	25		
WH2	HB278			21	27
WH2	HB82			23	28
WH3	HB7	17.5	22		
WH2	HB104				约39
WH3	HB132			27	31

腹内:齿板显著,短、粗,延伸在腹肌痕面两后侧;腹肌痕面显著,肌痕围脊强,肌痕面形状多变,呈亚心脏形或亚三角形;闭肌痕居中,长条状,但其前端短于其两侧的启肌痕;启肌痕相对较大,呈月牙状,其前端包围着闭肌痕;无腹中隔板。

背内:铰窝小,呈短三角状;腕基短,耸立;腕基支板短、粗,约呈 58°～70°(多为 70°)夹角异向展伸至背肌痕面后方,并且很少接触到背肌痕面;背窗台纵中部显著隆起,并与低、宽背中肌隔后端相融合;主突起小,椭圆状,位于背窗台后部,其冠部有时具有小锯片状横板,其茎部有时向前延伸成细脊状;背肌痕面很显著,方圆形,被低、宽中肌隔和横肌隔分成前后两对,而且前对闭肌痕远大于后对闭肌痕(图版 35,图 5,6,8,9,11,12;插图 25-B);背中肌隔低、宽,并限制在背肌痕面内。

比较:*Hirnantia sagittifera*(M'Coy)morph. Bohemia Havliček(1977)是以其前对闭肌痕远大于后对闭肌痕这一重要特征与 *Hirnantia* 其他各个种相区别。

产地层位:湖北宜昌王家湾、丁家坡;上奥陶统顶部五峰组观音桥段(赫南特阶中部)。

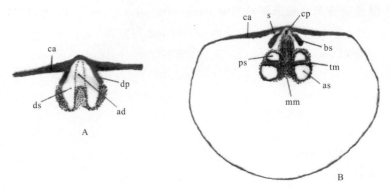

插图 25 *Hirnantia sagittifera* morph. Bohemia 的内部构造

Illustr. 25 Interior structures of *Hirnantia sagittifera* morph. Bohemia

A-腹内模(据图版 35,图 1,2);B-背内模(据图版 35,图 6,9)

A - Ventral internal mold(from pl. 35, figs. 1,2); B - Dorsal internal mold(from pl. 35, figs. 6,9)

ad-闭肌痕(adductor scars); as-前闭肌痕(anterior adductor scars); bs-腕基支板(brachiophore support); ca-铰合面(cardinal area); cp-主突起(cardinal process); dp-齿板(dental plate); ds-启肌痕(didductor scars); mm-中肌隔(median myophragm); ps-后闭肌痕(posterior adductor scars); tm-横肌隔(transverse myophragm); s-铰窝(socket)

箭形赫南特贝波兰形态种 *Hirnantia sagittifera* (M'Coy,1851) morph. Poland Temple(1965)

图版(pl.)37,图(figs.)1-11;插图(Illustr.)26

1965 *Hirnantia sagittifera* (M'Coy); Temple. P. 394, pl. 8, figs. 1-10; pl. 9, figs. 1-8.
2000 *Hirnantia sagittifera* (M'Coy); Harper. P. 826, fig. 603, 1f, 1g.

特征简要:贝体小,通常壳长4.1~5.3mm,壳宽5~6.9mm(表30);轮廓横亚椭圆形或亚圆形;铰合线直,短于最大壳宽;主端钝圆或钝角状。腹壳凸度中等;顶区凸度较强;腹铰合面适度高,斜倾型。背壳缓凸,接近于平坦,表面纵中部微凹成浅宽背中槽,但有时不明显;背铰合面低,正倾型。壳表饰较粗放射纹,通常作1~2次分枝。疹壳。

表30 箭形赫南特贝波兰形态种介壳测量(单位:mm)
Table 30 Shell measurements of *Hirnantia sagittifera* (M'Coy) morph. Poland Temple (in mm)

| 采集号 | 登记号 | 腹壳(ventral valve) | | 背壳(dorsal valve) | |
(Coll. No.)	(Cat. No.)	长(length)	宽(width)	长(length)	宽(width)
DH2	HB204	5	5.8		
WH1	HB200			3.6	4.7
DH3	HB507	5.3	6.4		
DH3	HB301	3	4		
DH3	HB396	5.3	6.9		
WH1	HB247	5.2	6.2		
DH2	HB714			4.5	约4.7
WH2	HB199			4.5	4.8
DH3	HB615			4.1	5
WH3	HB649			4.2	5.1
WH2	HB502			3.3	4.2

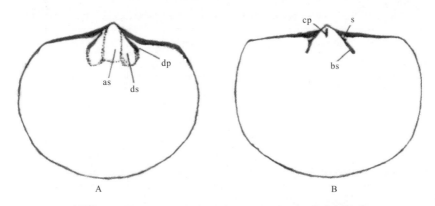

插图 26　*Hirnantia sagittifera* morph. Poland 的内部构造
Illustr. 26　Interior structures of *Hirnantia sagittifera* morph. Poland
A-腹内模(据图版 37,图 5);B-背内模(据图版 37,图 8)
A - Ventral internal mold(from pl. 37,fig. 5);B - Dorsal internal mold(from pl. 37,fig. 8).
as-闭肌痕(adductor scars);bs-腕基支板(brachiophore support);cp-主突起(cardinal process);
dp-齿板(dental plate);ds-启肌痕(didductor scars);s-铰窝(socket)

腹内:齿板短、粗,异向展伸在腹肌痕面两后侧;腹肌痕面大,亚三角形;闭肌痕居中,近窄长方形,其前缘近于平直,但前端稍微短于其两侧的启肌痕;启肌痕面较小,呈半月状,其前端轻微包围着居中的闭肌痕。

背内:铰窝小,短三角状;铰窝底板薄弱;腕基短、粗,耸立;腕基支板发育,约呈 48°～68°夹角异向展伸;背窗腔宽阔,三角状;主突起呈细脊状,位于背窗腔后部,其冠部有时具有锯齿状小横片(图版 37,图 9b),其茎部有时轻微向前延伸成短细脊状;背肌痕面不清晰。

比较:*Hirnantia sagittifera*(M'Coy)morph. Poland Temple 的主要特征是贝体小,背壳凸度较平缓,腕基支板薄,较长,背窗腔较宽阔,背肌痕面不显露等特征与 *Hirnantia* 其他各个种明显可分。

产地层位:湖北宜昌王家湾、丁家坡;上奥陶统顶部五峰组观音桥段(赫南特阶中部)。

大赫南特贝 *Hirnantia magna* Rong,Xu et Yang
图版(pl.)36,图(figs.)3-12;插图(Illustr.)27

1974　*Hirnantia magna* Rong,Xu et Yang. 196 页,图版 92,图 25,26。
1982　*Hirnantia sagittifera* M'Coy;Fu. 106 页,图版 32,图 8,9。
1987　*Hirnantia magna* Rong,Xu et Yang;Zeng. 222 页,图版 12,图 1-3。

特征简要:贝体大,最大壳宽可达 43mm(表 31);轮廓横椭圆形;侧视近等双凸型;铰合线直,短于最大壳宽;最大壳宽位于贝体横中部;主端钝圆或近钝角状。腹壳凸度低,顶区凸度较强;腹铰合面低,斜倾型。背壳缓凸,顶区稍微隆起,纵中线有时微凹,但不足为背中槽;背铰合面低,正倾型。壳表放射纹细密,纹顶圆滑。

腹内:齿板短、粗,延伸在腹肌痕面两后侧;腹肌痕面清晰,肌痕围脊显著,但肌痕形状多变,并且以三角状居多;闭肌痕居中,窄长;启肌痕较大,半月形,其前端稍长于闭肌痕。

背内:铰窝小,短三角状,铰窝底板较发育;腕基极短,耸立;腕基支板短、粗,呈 60°～65°夹角异向展伸至背肌痕面后方,但有时接触到背肌痕面后缘;背窗腔相对较短小,呈三角状;主突起小,位于背窗腔后部,冠部有时具有锯片状小横板;背肌痕面清晰,被中肌隔和横肌隔分成前后两对,而且后对闭肌痕明显大于前对闭肌痕(图版 36,图 5-9,11,12;插图 27)。

比较:*Hirnantia magna* 的最大特点是贝体大,背内的后对闭肌痕面远大于前对闭肌痕面,并且以此与 *Hirnantia* 其他各个种相区别。

表 31 大赫南特贝介壳测量(单位:mm)
Table 31 Shell measurements of *Hirnantia magna* Rong, Xu et Yang (in mm)

采集号 (Coll. No.)	登记号 (Cat. No.)	腹壳(ventral valve)		背壳(dorsal valve)	
		长(length)	宽(width)	长(length)	宽(width)
WH3	HB8	24.8	28		
WH2	HB7	20	24		
WH2	HB976			约 30	约 40
DH2	HB59			32.5	约 43
WH3	HB128				约 16
WH3	HB119			23.5	32.5
WH3	HB65	32	38		
WH3	HB296			26	29

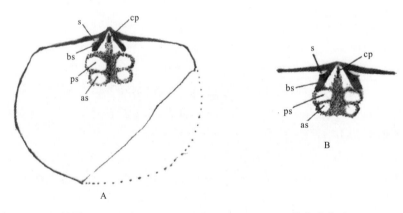

插图 27 *Hirnantia magna* Rong, Xu et Yang 的背内构造

Illustr. 27 Interior structures of dorsal valves of *Hirnantia magna* Rong, Xu et Yang

A-背内模(据图版 36,图 7);B-背内模(据图版 36,图 5)

A - Dorsal internal mold(from pl. 36, fig. 7); B - Dorsal internal mold(from pl. 36, fig. 5)

as -前闭肌痕(anterior adductor scars); bs -腕基支板(brachiophore support); cp -主突起(cardinal process); ps -后闭肌痕(posterior adductor scars); s -铰窝(socket)

产地层位:湖北宜昌王家湾、丁家坡;上奥陶统顶部五峰组观音桥段(赫南特阶中部)。

隔板赫南特贝 *Hirnantia septumis* Zeng

图版(pl.)38,图(figs.)1-12;插图(Illustr.)28

1987 *Hirnantia septa* Zeng.223 页,图版 12,图 11-14。
1987 *Hirnantia trilobata* Zeng.223 页,图版 12,图 9,10。

特征简要:贝体中等至大,通常壳长 20~30mm,壳宽 23~36mm(表 32);轮廓亚圆;侧视近等双凸型;铰合线直,短于最大壳宽,最大壳宽位于贝体横中部;主端钝圆。腹壳缓凸,在顶区稍微隆起;腹铰合面低,斜倾型。背壳凸度低,仅在顶区凸度较强;背铰合面低,正倾型。壳表放射纹细密,纹顶圆滑,作 2~3 次分枝。

表 32　隔板赫南特贝介壳测量(单位:mm)

Table 32　Shell measurements of *Hirnantia septumis* Zeng(in mm)

采集号 (Coll. No.)	登记号 (Cat. No.)	腹壳(ventral valve)		背壳(dorsal valve)	
		长(length)	宽(width)	长(length)	宽(width)
WH1	HB85	22.2	24.5		
花 H2	HB746	21.3	24		
DH2	HB84	22	24.3		
HH2	HB748	20	23		
WH2	HB80	14	15.5		
HH2	HB749	22	约 25		
WH3	HB56			27	34
HH2	HB747			18.7	24.6
WH1	HB13			约 30	36
WH3	HB486			25	26.5

腹内:齿板强,轻微向内弯,延伸至腹肌痕面两后半侧;腹肌痕面显著,肌痕围脊强,呈亚三角形或者亚卵形;闭肌痕面居中,呈窄长条状,其前端稍短于两侧的启肌痕;启肌痕面较大,呈月牙状;腹肌痕之前具细中隔板(图版 38,图 1-8;插图 28-A)。

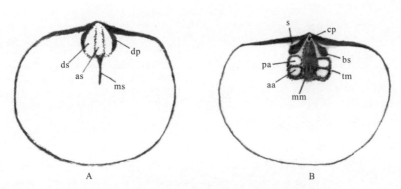

插图 28　*Hirnantia septumis* Zeng 的内部构造

Illustr. 28　Interior structures of *Hirnantia septumis* Zeng

A-腹内模(据图版 38,图 1,3);B-背内模(据图版 38,图 9,10)

A-Ventral internal mold(from pl. 38,figs.1,3);B-Dorsal internal mold(from pl. 38,figs.9,10).

aa-前闭肌痕(anterior adductor scars);as-闭肌痕(adductor scars);bs-腕基支板(brachiophore support);cp-主突起(cardinal process);dp-齿板(dental plate);ds-启肌痕(didductor scars);mm-中肌隔(median myophragm);ms-中隔板(median septum);pa-后闭肌痕(posterior adductor scars);s-铰窝(socket);tm-横肌隔(transverse myophragm)

背内:铰窝小,长三角状;腕基极短,耸立;腕基支板短、粗,呈 60°~70°夹角异向展伸至背肌痕面后方;背窗台纵中部强烈隆起呈宽圆脊状,其前端与中肌隔融合;主突起小,位于背窗台后部;背肌痕显著,呈近方形或方圆形,并且被粗宽中肌隔和横肌隔分成近等大的前后两对(图版 38,图 9-12;插图 28-B);粗、宽中肌隔多数限在背肌痕面内,但有的穿越背肌痕面前缘,形成粗背中隔脊(图版 38,图 11)。

比较:*Hirnantia septumis* 的最主要特点是具细腹中隔板,背内前后两对闭肌痕近等大,介于 *H. sagittifera* morph. Bohemia 和 *H. magna* 两个种之间的一种背肌痕面形态。

产地层位:湖北宜昌王家湾、丁家坡;上奥陶统顶部五峰组观音桥段(赫南特阶中部)。

丰富赫南特贝 *Hirnantia fecunda* Rong

图版(pl.)34,图(figs.)6-12

1979 *Hirnantia sagittifera fecunda* Rong.2 页,图版 1,图 10,14,15。
1983 *Hirnantia sagittifera fecunda* Rong;Zeng.118 页,图版 14,图 5-8。

特征简述:贝体中等至大,通常壳长 11.2~21mm,壳宽 14.5~25.5mm(表 33);轮廓亚圆形;侧视低双凸型;铰合线直,短于最大壳宽,最大壳宽位于贝体横中部;主端钝圆。腹壳缓凸,顶区凸度稍强,纵中部呈轻微龙脊状;腹铰合面低,斜倾型。背壳凸度低,接近于平坦,仅在顶区稍微隆起,纵中部微凹,但不足为背中槽;背铰合面低,正倾型。壳表饰细密放射纹,纹顶圆滑,作 2~3 次分枝,有时局部呈簇型放射纹。疹壳。

表 33 丰富赫南特贝介壳测量(单位:mm)
Table 33 Shell measurements of *Hirnantia fecunda* Rong(in mm)

采集号 (Coll. No.)	登记号 (Cat. No.)	腹壳(ventral valve)		背壳(dorsal valve)	
		长(length)	宽(width)	长(length)	宽(width)
WH1	HB46			21	25.5
DH3	HB712			14.2	17
DH2	HB553			9.5	10
DH3	HB514			12.3	15
DH2	HB1	12.5	16.4		
WH2	HB464			11.2	14.5
DH3	HB526			18.5	22

腹内:齿板短、粗,延伸在腹肌痕面两后侧;腹肌痕面显著,呈亚三角形;闭肌痕居中,呈长方形,其前端稍短于两侧的启肌痕;启肌痕较大,呈近半月状。

背内:铰窝极浅小;铰窝底板发育;腕基极短小,呈小三角锥状;腕基支板薄板状,相对较长,多数为 38°~48°夹角异向展伸;主突起微小,椭圆形,位于背窗腔后部,茎部延伸出极弱细脊;背肌痕不清晰。

讨论:在宜昌地区观音桥段中经常见到一些贝体的大小、外部形态以及背内的特征都与 *Hirnantia sagittifera fecunda* Rong 雷同的标本。尤其是它们的铰窝极小;铰窝底板发育;腕基极小;腕基支板薄板状,但又相对较长,异向展伸的夹角都较小,背肌痕面不清晰等特征更是与 *Hirnantia sagittifera fecunda* 无差别。因此当前的标本应为 *H. sagittifera fecunda* Rong。又因为这些标本的腕基极短小,尤其是腕基支板却相对较长,异向展伸的夹角多数为 38°~48°,以及其背肌痕面不清晰等特征而与 *Hirnantia* Lamont(1935)已知的各个种有明显的差别,应作为独立的一个种。因此将当前的标本做出前面的鉴别。

产地层位:湖北宜昌王家湾、丁家坡;上奥陶统顶部五峰组观音桥段(赫南特阶中部)。

辛奈贝属 Genus *Kinnella* Bergström,1968

1968 *Kinnella* Bergström. P.11.
1976 *Kinnella* Bergström;Lesperance et Sheehan. P.724.
1977 *Kinnella* Bergström;Havliček. P.269.
1978 *Kinnella* Bergström;Yan.216 页。
1979 *Kinnella* Bergström;Rong.2 页。
1981 *Kinnella* Bergström;Chang.560 页。

| 1982 | *Kinnella* Bergström；Fu. 107 页。
| 1983 | *Kinnella* Bergström；Zeng. 116 页。
| 1984 | *Kinnella* Bergström；Rong. P. 142.
| 1987 | *Kinnella* Bergström；Zeng. 223 页。
| 2000 | *Kinnella* Bergström；Harper. P. 831.
| 2006 | *Kinnella* Bergström；Rong. 294 页。

属型种：Genotype *Hirnantia*？*kielanae* Temple，1965.

特征简要：贝体小，轮廓亚圆形，侧视强烈腹双凸型；铰合线直，稍微短于最大壳宽；主端圆滑状。腹壳强凸，最大凸度位于喙部附近；腹喙大，尖突，成为腹壳最高点，也是腹铰合面的最高点；腹铰合面高强，下倾型，甚至接近于前倾型；腹三角孔狭窄，长三角状，洞开。背壳缓凸，成年壳纵中线凹下形成浅宽背中槽；背铰合面中等高，正倾型；背三角孔洞开。壳表饰放射纹。疹壳。

腹内：齿板短粗，延伸在腹肌痕面两后侧；腹肌痕面亚圆形；具显著的肌痕围脊。

背内：铰窝显著，长三角形，呈宽"八"字形展布；腕基短粗；腕基支板短、粗壮，小角度异向或近平行展伸；主突起显著，双叶状，锯齿状小横板（Crenulations）发育；背肌痕面大，方圆形，被背中隔脊和横肌隔分成前、后两对，而且前对闭肌痕远大于后对闭肌痕；背中隔脊发育，其后端与主突起前端相融合。

分布及时代：中国中南部、欧洲；晚奥陶世赫南特期中期。

基兰辛奈贝 *Kinnella kielanae* (Temple)

图版(pl.)39，图(figs.)1-10；图版(pl.)40，图(figs.)1-8；

图版(pl.)51，图(fig.)2；插图(Illustr.)29-A

| 1965 | *Hirnantia*？*kielanae* Temple. P. 401, pl. 8, figs. 1, 3, 4; pl. 9, figs. 1-8; pl. 10, figs. 1-8; pl. 11, figs. 1-7.
| 1967 | *Hirnantia kielanae* Temple；Marek et Havlíček. P. 280, pl. 2, figs. 5, 6.
| 1968 | *Kinnella kielanae* (Temple)；Bergström. P. 11, pl. 4, figs. 3-6.
| 1975 | *Hirnantia kielanae* Temple；Fu. 111 页，图版 23，图 5-7。
| 1976 | *Kinnella kielanae* (Temple)；Lesperance et Sheehan. P. 724, pl. 109, figs. 12-19.
| 1977 | *Kinnella kielanae proclinis* Havlíček. P. 270, pl. 30, figs. 5-18.
| 1978 | *Kinnella kielanae* (Temple)；Yan. 216 页，图版 62，图 20-22。
| 1979 | *Kinnella kielanae* (Temple)；Rong. 2 页，图版 1，图 6, 8, 9, 11。
| 1982 | *Kinnella kielanae* (Temple)；Fu. 107 页，图版 32，图 10-12。
| 1983 | *Kinnella kielanae* (Temple)；Zeng. 116 页，图版 15，图 6-13。
| 1987 | *Kinnella kielanae* (Temple)；Zeng. 223 页，图版 12，图 15-18。
| 2000 | *Kinnella kielanae* (Temple)；Harper. P. 831, figs. 604, 1a, 1f.

描述：贝体小，通常壳长 3～5.5mm，壳宽 4.2～6.8mm（表 34）；轮廓亚圆形或横椭圆形；侧视强烈腹双凸型；铰合线直，稍微短于最大壳宽，最大壳宽位于贝体横中部；主端钝圆；成年体前接合缘轻微单槽型。腹壳强凸，最大凸度位于腹喙部附近；腹喙大，耸立，成为腹壳最高点，也是腹铰合面最高点；腹铰合面高，强烈斜倾型或下倾型；腹三角孔呈狭窄三角形，洞开。背壳缓凸，成年体纵中线微凹形成宽浅背中槽；背铰合面中等高，正倾型；背三角孔洞开。壳表饰放射纹，纹顶圆脊状，作 1～2 次分枝。疹壳。

腹内：齿板短粗，延伸在腹肌痕面两后侧；腹肌痕面显著，方圆形，不易区分；肌痕面围脊发育。

背内：铰窝显著，长三角形，呈宽"八"字形展布；腕基短粗，耸立；腕基支板短，粗壮，呈 30°～38°夹角异向展伸至背肌痕面后方；主突起呈长椭圆形，双叶型，锯齿状小横板（crenulations）发育，主突起前端与粗强背中隔脊后端相融合（图版 40，图 6a，4b；插图 29-A）；背肌痕面显著，近方圆形，被背中隔脊和横肌隔分成前后两对；前对闭肌痕远大于后对闭肌痕，呈卵圆形；后对闭肌痕较小，近圆形，其后缘圆滑（图版 39，图 4；图版 40，图 4a；插图 29-A）；背中隔脊粗强，从主突起前缘直伸至背肌痕面前缘，有时穿越背肌痕面前缘。

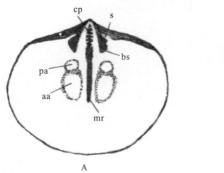
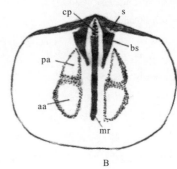

插图 29 背内构造对比图

Illustr. 29 The correlative figures of dorsal interior structures

A - *Kinnella kielanae* (Temple) 的背内模 (据图版 40, 图 4a, 4b); B - *Kinnella robusta* Chang 的背内模 (据图版 41, 图 3, 9)

A - The dorsal internal mold of *Kinnella kielanae* (Temple) (from pl. 40, figs. 4a, 4b); B - The dorsal internal mold of *Kinnella robusta* Chang (from pl. 41, figs. 3, 9).

aa - 前闭肌痕 (anterior adductor scars); bs - 腕基支板 (brachiophore support); cp - 主突起 (cardinal process); mr - 中隔脊 (median ridge); pa - 后闭肌痕 (posterior adductor scars); s - 铰窝 (socket)

表 34 基兰辛奈贝介壳测量 (单位: mm)

Table 34 Shell measurements of *Kinnella kielanae* (Temple) (in mm)

采集号 (Coll. No.)	登记号 (Cat. No.)	腹壳 (ventral valve)		背壳 (dorsal valve)	
		长 (length)	宽 (width)	长 (length)	宽 (width)
WH2	HB268	3	4.2		
WH1	HB262	3.3	5		
DH3	HB306			3	4
DH3	HB527			5.5	6.8
WH1	HB264			3.2	3.9
WH3	HB571			4.3	5.5
DH2	HB401			3.9	4.8
WH3	HB273			5.1	6.1
WH2	HB32			4	5
WH1	HB258	4.2	5.6		
WH3	HB242	3.6	5.3		
WH2	HB463	4	6.1		
DH3	HB368			4	5.2
WH3	HB259			5	6.9
WH1	HB266			3	4.2
WH1	HB263			2.7	3.8
WH1	HB265			3.8	4.7

产地层位:湖北宜昌王家湾、丁家坡;上奥陶统顶部五峰组观音桥段(赫南特阶中部)。

隆凸辛奈贝 *Kinnella robusta* Chang

图版(pl.)41,图(figs.)1-11;插图(Illustr.)29-B

1965 *Hirnantia? kielanae* Temple. P. 401,pl. 8,figs. 2,6,7.
1981 *Kinnella robusta* Chang. 560 页,图版 1,图 18-20。
1983 *Kinnella kielanae*(Temple);Zeng. 116 页,图版 15,图 8,11。
1984 *Kinnella kielanae*(Temple);Rong. P. 142,pl. 8,fig. 6.
2006 *Kinnella kielanae*(Temple);Rong. 294 页,图版 2,图 1,2,5,6。

讨论:*Kinnella robusta* Chang 这一个种是由常美丽于1981年建立的,标本是采集于宜昌黄花场上奥陶统五峰组观音桥段(或称观音桥层)。戎嘉余于1984年将 *K. robusta* Chang 视为 *Kinnella kielanae*(Temple)的同义名(P.143)。当前从宜昌王家湾、丁家坡两剖面的观音桥段采集到较多该类型,从保存较好的背内模标本来看,*K. robusta* Chang 这一个种应该存在。其贝体大小详见表35。*K. robusta* 与 *K. kielanae*(Temple)的主要区别是 *K. robusta* 背内的腕基支板较长,其前部轻微相向延伸,后对背肌痕面呈尖三角形,而且其后部包围着腕基支板的前部(图版41,图3,4a,6,8,9;插图29-B);而 *Kinnella kielanae*(Temple)背内的腕基支板较短,仅伸至背肌痕面的后方,而且后对背肌痕面呈亚圆形(图版39,图4;图版40,图4a;插图29-A)。因此,*K. robusta* Chang(1981)应当为独立的一个种。本种的背内构造与 *Draborthis* Marek et Havliček(1967)的背内构造非常接近,表明两者的关系非常密切。

表35 隆凸辛奈贝介壳测量(单位:mm)
Table 35 Shell measurements of *Kinnella robusta* Chang(in mm)

采集号 (Coll. No.)	登记号 (Cat. No.)	腹壳(ventral valve)		背壳(dorsal valve)	
		长(length)	宽(width)	长(length)	宽(width)
WH2	HB194	3	4.2		
WH2	HB269	3.2	4.7		
DH2	HB494			5	6.2
WH2	HB179			3.4	4
WH2	HB257			4	5.2
WH3	HB275			4.4	5.8
DH2	HB270			4.8	5.7
DH2	HB326			4.1	5.2
WH3	HB274			4.7	5.7
DH3	HB300			4.4	5.9
WH2	HB276			3.9	5

产地层位:湖北宜昌王家湾、丁家坡;上奥陶统顶部五峰组观音桥段(赫南特阶中部)。

德拉勃正形贝属 Genus *Draborthis* Marek et Havliček,1967

1967 *Draborthis* Marek et Havliček. P. 280.
1968 *Draborthis* Marek et Havliček;Bergström. P. 10.
1968 *Draborthis* Marek et Havliček;Temple. P. 44.

1975　*Draborthis* Marek et Havlíček；Fu. 112 页。
1977　*Draborthis* Marek et Havlíček；Havlíček. P. 272.
1979　*Draborthis* Marek et Havlíček；Rong. 2 页。
1981　*Draborthis* Marek et Havlíček；Chang. 560 页。
1983　*Draborthis* Marek et Havlíček；Zeng. 116 页。
1984　*Draborthis* Marek et Havlíček；Rong. P. 144.
2006　*Draborthis* Marek et Havlíček；Rong. 294 页。

属型种：Genotype *Draborthis caelebs* Marek et Havlíček，1967.

特征简要：贝体中等大，轮廓亚圆形，侧视近平凸型；铰合线直，短于最大壳宽；主端钝圆。腹壳缓凸，仅在顶区缓和隆起；腹喙大，微弯；腹铰合面中等发育，斜倾型；腹三角孔洞开。背壳凸度平缓；背喙极短小，轻微突向背方；背铰合面低，强烈正倾型或轻微超倾型；背三角孔洞开。壳表饰较粗的放射纹，纹顶具有前倾的中空壳刺。疹壳。

腹内：齿板较长，异向或者近平行延伸在腹肌痕面两外侧；腹肌痕面显著，亚卵形或近纵长方形；闭肌痕面居中，较大，其前端稍微长于其两侧的启肌痕；启肌痕较小，呈窄月牙状。

背内：主基短而宽阔。铰窝显著，呈窄三角形；内铰窝脊（腕基）长，与铰合线近平行或者强烈异向展布（图版 42，图 6，12）；腕基支板发育，相向延伸在背中隔脊两侧；背窗腔短，后部较深凹；主突起显著，冠部呈弱双叶型，锯齿状小横板弱小，主突起茎部前端与背中隔脊后端相融合；背肌痕面显著，方圆形，被背中隔脊和微弱横肌隔分为前后两对；前对闭肌痕较大，近卵圆形；后对闭肌痕较小，呈尖三角形，并包围着大部分腕基支板（图版 42，图 6；图版 43，图 7，8）；背中隔脊粗，从主突起茎部前端延伸至背肌痕面前缘。

比较：*Draborthis* 在 Draboviidae 科内是很特殊的一个属。它的放射纹上具有向前倾伏的中空壳刺；背铰合面低，强烈正倾型或轻微超倾型；齿板较长，近平行延伸在亚卵形腹肌痕两侧；腕基与铰合线近平行；腕基支板相向延伸在背中隔脊两侧；后对闭肌痕面呈尖三角形，并且几乎包围着腕基支板；主突起冠部具微弱锯齿状小横板，其茎部前端与背中隔脊后端相融合等特征与 Draboviidae 科内的各个属明显不同。

分布及时代：中国中南部、欧洲；晚奥陶世赫南特期中期。

孤独德拉勃正形贝 *Draborthis caelebs* Marek et Havlíček

图版（pl.）42，图（figs.）1–12；图版（pl.）43，图（figs.）1–11；
插图（Illustr.）30

1967　*Draborthis caelebs* Marek et Havlíček. P. 280，pl. 2，figs. 9，10，16，17.
1968　*Draborthis caelebs* Marek et Havlíček；Bergström. P. 10，pl. 3，figs. 5–7.
1968　*Draborthis* cf. *caelebs* Marek et Havlíček；Temple. P. 44，pl. 8，figs. 1–8.
1975　*Draborthis caelebs* Marek et Havlíček；Fu. 112 页，图版 23，图 8，9。
1977　*Draborthis caelebs* Marek et Havlíček；Havlíček. P. 272，pl. 31，figs. 11–18.
1979　*Draborthis caelebs* Marek et Havlíček；Rong. 2 页，图版 1，图 12。
1983　*Draborthis caelebs* Marek et Havlíček；Zeng. 116 页，图版 15，图 1–5。
1984　*Draborthis caelebs* Marek et Havlíček；Rong. P. 145，pl. 9，figs. 1–5，7–10，12–14.
2006　*Draborthis caelebs* Marek et Havlíček；Rong. 294 页，图版 2，图 11。

描述：贝体中等大，通常壳长 6～11.6mm，壳宽 7～13mm（表 36）；轮廓亚圆形；侧视近平凸型；铰合线直，短于最大壳宽，最大壳宽位于贝体横中部；主端近钝圆；侧缘和前缘近于圆弧状。腹壳缓凸，顶区凸度较强；腹喙大，轻微向背方压下；腹铰合面中等高，斜倾型；腹三角孔洞开。背壳凸度低，近于平坦；背喙很小，轻微突向背壳顶；背铰合面较低，强烈正倾型或者轻微超倾型；背三角孔洞开。壳表饰较粗的放射纹；纹顶具有向前倾伏的中空壳刺（图版 42，图 4，7）；同心纹稀少。

表36 孤独德拉勃正形贝介壳测量(单位:mm)
Table 36 Shell measurements of *Draborthis caelebs* Marek et Havlíček (in mm)

采集号 (Coll. No.)	登记号 (Cat. No.)	腹壳(ventral valve)		背壳(dorsal valve)	
		长(length)	宽(width)	长(length)	宽(width)
DH2	HB547	6.8	7.2		
DH2	HB474	7.5	11.3		
DH2	HB475	6	7		
DH3	HB479	10.3	12.7		
DH3	HB108				9.5
HH3(=HK3)	HB753			8	10.3
DH3	HB530	8.6	8		
WH3	HB476			6	7.6
WH3	HB704			8.1	10.3
DH3	HB478			8.1	11.2
DH3	HB297			6.8	8.5
WH2	HB397			8.6	10.8
DH3	HB109	9.5	10.5		
DH2	HB320	7	11.3		
DH2	HB205	11.6	13		
HH3(=HK3)	HB736			8.3	11
DH3	HB152			7.5	10.2
DH3	HB645			10.2	12
DH3	HB522			6.5	7.6
DH3	HB348			10.1	11.9
DH3	HB477			5	8
WH2	HB472			5.2	7

腹内:齿板薄,较长,轻微异向展伸或近平行延伸在腹肌痕两侧;腹肌痕面清晰,呈亚卵形;闭肌痕面居中,较大,其前端稍微长于其两侧的启肌痕;启肌痕面较小,呈窄月牙形。

背内:铰窝显著,窄三角形(图版42,图5;图版43,图10);腕基薄板状,强烈异向展伸,几乎与铰合线平行(图版42,图6,12;图版43,图8),但有时较短粗;腕基支板发育,相向向前延伸于背中隔脊两侧边,有时其前端几乎与背中隔脊相会合;背窗腔短,后部较为深凹;主突起显著,窄椭圆状,冠部具微弱锯状小横板(图版42,图11;图版43,图11b),主突起茎部前端与背中隔脊后端相融合;背肌痕面大,方圆形或长椭圆形,并且被背中隔脊和微弱横肌隔分成前后两对;前对闭肌痕面较大,呈椭圆形;后对闭肌痕面较小,呈尖三角形,并且几乎包围着腕基支板(图版42,图5,6;图版43,图5,7,8;插图30),但有时前后对闭肌痕不易区分(图版43,图4);背

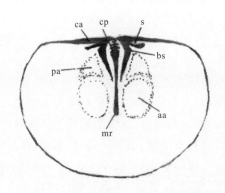

插图30 *Draborthis caelebs* 的背内模
(据图版42,图5,6)

Illustr. 30 Dorsal internal mold of *Draborthis caelebs* (from pl. 42, figs. 5, 6)

aa-前闭肌痕(anterior adductor scars);bs-腕基支板(brachiophore support);ca-铰合面(cardinal area);cp-主突起(cardinal process);mr-中隔脊(median ridge);pa-后闭肌痕(posterior adductor scars);s-铰窝(socket)

中隔脊发育,但其前端不穿越背肌痕面的前缘。

产地层位:湖北宜昌王家湾、丁家坡;上奥陶统顶部五峰组观音桥段(赫南特阶中部)。

难得正形贝科(新科)Family Dysprosorthidae Zeng et Zhang(fam. nov.)

模式属:Type genus *Dysprosorthis* Rong,1984.

词源:建立在 *Dysprosorthis* 模式属的基础上。

特征:贝体小,近等双凸型;铰合线直,短于最大壳宽;铰合面窄,而且很低;轮廓亚圆形;壳表饰典型簇型放射纹,近前缘处具有少量叠瓦状同心层。疹壳,疹管(盲孔)很细、稀疏、放射状排列在壳线内。

腹内:齿板薄板状,近平行延伸在腹肌痕面两外侧;腹肌痕面呈亚卵形。

背内:主基开阔,但很矮;铰窝窄坑状,位于铰合线之后,并呈宽"八"字形展布在背三角孔两前半侧;铰窝外边缘与背三角孔后侧边斜交;腕基(内铰窝脊)粗脊状,与背三角孔前侧边近平行;背窗腔短,但很开阔,呈三角状;主突起单刃状。

讨论:新科 Dysprosorthidae(fam. nov.)的主基非常独特,尤其是它的铰窝位于铰合线之后,并呈宽"八"字形展布在背三角孔两前半侧;其外边缘(外铰窝脊)与背三角孔后侧边斜交;腕基(内铰窝脊)呈粗棱脊状,并与背三角孔前侧边近平行;背窗腔短,但很开阔,呈三角状;主突起单刃状。这些不仅是违反常规、而且是很不合理的一种主基形态,在德姆贝亚目 Suborder Dalmanellidina Moore(1952)中是从未见过的,它与 Family Chrustenoporidae Havlíček et Mergl(1982)众成员的性质截然不同。因此不能把 *Dysprosorthis* Rong(1984)归入到 Chrustenoporidae 这个科内(Harper,2000,P.836),而应以 *Dysprosorthis* Rong(1984)为模式属,另建立 Dysprosorthidae(fam. nov.)这一新科。

时代:晚奥陶世赫南特期中期(Middle Hirnantian)。

难得正形贝属 Genus *Dysprosorthis* Rong,1984

1984 *Dysprosorthis* Rong;P.133.
2000 *Dysprosorthis* Rong;Harper. P.836.
2006 *Dysprosorthis* Rong;Rong. 294 页。

属型种:Genotype *Dysprosorthis sinensis* Rong,1984.

特征简要:贝体小,轮廓亚圆形,侧视近等双凸型;铰合线直,短于最大壳宽;主端钝圆或钝角状;两侧缘和前缘圆弧状;前接合缘轻微单槽型。腹壳缓凸,顶区凸度相对较大;腹铰合面矮小,斜倾型;腹三角孔洞开;腹中隆呈龙骨状,是由簇型壳纹隆起而形成,始于喙部之前,直伸至前缘。背壳缓凸;背铰合面矮小,强烈正倾型;背三角孔短而宽,洞开;背中槽浅宽,始于喙部前方,向前逐步加宽直至前缘。壳表饰典型的簇型壳纹,成年体近前缘处具有少量叠瓦状同心层。疹壳。

腹内:齿板薄板状,近平行延伸在肌痕面两外侧;腹肌痕面亚卵形,其前缘具发育程度不同的围脊;闭肌痕居中,长椭圆形,其前端长于其两侧的启肌痕;启肌痕相对较短,呈长月牙形。

背内补充描述:主基开阔,但很矮,很特殊。铰窝细坑状,位于铰合线之后,并呈宽"八"字形展布在背三角孔两前半侧(图版44,图2b,4a,4b;图版55,图1,2);铰窝外边缘与背三角孔后侧边斜交;腕基(内铰窝脊)粗棱脊状,与背三角孔两前侧边近平行;无腕基支板,但在老年壳可见极为短小、雏形的腕基支板(图版44,图9);背窗腔短而宽,主突起单刃状;铰窝底板厚而宽(从背内模后斜视,在铰窝内模下方可见到被溶蚀后的板状空间);背肌痕未显露。

分布及时代:中国中南部、英格兰?爱尔兰?摩洛哥?;晚奥陶世赫南特期中期(Middle Hirnantian)。

中华难得正形贝 *Dysprosorthis sinensis* Rong

图版(pl.)44,图(figs.)1-9;图版(pl.)55,图(figs.)1,2;

插图(Illustr.)31-A,B

1984 *Dysprosorthis sinensis* Rong. P. 135,pl. 4,2,5,10?;pl. 5,figs. 1-13;text-fig. 9.
2000 *Dysprosorthis sinensis* Rong;Harper. P. 836,fig. 607,2a,2b.
2006 *Dysprosorthis sinensis* Rong;Rong. 294页,图版2,图9,10。

描述:贝体小,除少年期贝体外,通常壳长4.2～8.5mm,壳宽5～10.8mm,个别大者的壳长可达10.3mm,壳宽约达13.5mm(表37);轮廓亚圆形;侧视近等双凸型;铰合线直,短于最大壳宽,主端钝圆或钝角状;前接合缘轻微单槽型。腹壳缓凸,在顶区凸度较大;腹铰合面矮小,斜倾型;腹三角孔洞开;壳表纵中线具龙骨状腹中隆,从喙部前方直伸至前缘,是由3束簇型壳纹组成。背壳缓凸;背铰合面矮小,强烈正倾型;背三角孔短而宽,洞开;壳表纵中部具浅宽背中槽(图版44,图2a,4a,6),始于喙部前方,并向前逐步加宽直达前缘。壳表饰典型的簇型放射纹,在近前缘处具有少量叠瓦状同心层(图版44,图2a,6)。疹壳(图版44,图1,2a,4a)。

表37 中华难得正形贝介壳测量(单位:mm)

Table 37 Shell measurements of *Dysprosorthis sinensis* Rong(in mm)

采集号 (Coll. No.)	登记号 (Cat. No.)	腹壳(ventral valve)		背壳(dorsal valve)	
		长(length)	宽(width)	长(length)	宽(width)
WH3	HB602	4.2	5		
WH2	HB211			7.4	9
WH2	HB212			4.9	6.2
WH2	HB158			4.5	5.5
WH2	HB218			3.9	5.2
WH2	HB219			8.5	10.8
WH2	HB213			约3.2	4.2
DH2	HB705			2.3	3
WH1	HB15			10.3	约13.5

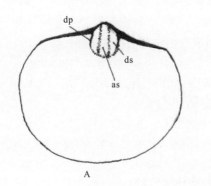
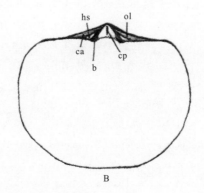

插图31 *Dysprosorthis sinensis* Rong 的内部构造

Illustr. 31 Interior structures of *Dysprosorthis sinensis* Rong

A-腹内模(据图版44,图1);B-背内模(据图版44,图2a,2b,4a,4b)

A-Ventral internal mold(from pl. 44,fig. 1);B-Dorsal internal mold(from pl. 44,figs. 2a,2b,4a,4b)

as-闭肌痕(adductor scars);b-腕基(brachiophore);ca-铰合面(cardinal area);cp-主突起(cardinal process);dp-齿板(dental plate);ds-启肌痕(didductor scars);hs-铰窝(hinge socket);ol-铰窝外边缘(outer socket limbus)

背、腹内构造同属征。

产地层位：湖北宜昌王家湾、丁家坡；上奥陶统顶部五峰组观音桥段（赫南特阶中部）。

小嘴贝目 Order Rhynchonellida Kuhn, 1949

　　孔嘴贝超科 Superfamily Rhynchotrematoidea Schuchert, 1913

　　　　三角嘴贝科 Family Trigonirhynchiidae Schmidt, 1965

　　　　　　嘴室贝亚科 Subfamily Rostricellulinae Rozman, 1969

小褶窗贝属 Genus *Plectothyrella* Temple, 1965

1965　*Plectothyrella* Temple. P. 412.
1967　*Plectothyrella* Temple; Marek et Havliček. P. 284.
1968　*Plectothyrella* Temple; Bergström. P. 19.
1974　*Plectothyrella* Temple; Rong. 199 页。
1975　*Plectothyrella* Temple; Fu. 113 页。
1977　*Plectothyrella* Temple; Zeng. 66 页。
1979　*Plectothyrella* Temple; Rong. 2 页。
1981　*Plectothyrella* Temple; Chang. 563 页。
1983　*Plectothyrella* Temple; Zeng. 121 页。
1984　*Plectothyrella* Temple; Rong. P. 162.
1999　*Plectothyrella* Temple; Villas, Lorenzo et Marco. P. 193.
2000　*Plectothyrella* Temple; Carlson et al. P. 1065.
2006　*Plectothyrella* Temple; Rong. 294 页。

属型种：Genotype *Plectothyrella platystrophoides* Temple(1965) = *P. crassicosta* (Dalman), 1828.

特征简要：贝体中等大，轮廓亚圆形或亚三角形；侧视强双凸型；铰合线短；主端阔圆形；背中隆和腹中槽都较微弱；腹喙较大，悬弯在背喙之上；腹三角孔洞开。壳表饰放射褶。

腹内：齿板发育，异向延伸在深凹腹肌痕面两外侧；腹肌痕面深凹，其两侧被齿板所限制。

背内：铰窝椭圆状；内铰窝脊短粗；腕基粗强，其顶端伸出很长的腕棒，弯曲，并且强烈向腹方翘起；铰板分离；隔板槽？短；背中隔脊长。

分布及时代：中国中南部、欧洲、北美东部、北非、南非、西伯利亚；晚奥陶世赫南特期中期。

粗线小褶窗贝 *Plectothyrella crassicosta* (Dalman)

图版(pl.)47, 图(figs.)7 – 9; 图版(pl.)48, 图(figs.)1 – 10;

图版(pl.)60, 图(figs.)1, 2; 插图(Illustr.)32 – A, B

1828　*Atrypa? crassicosta* Dalman; 1827. P. 131 – 132.
1965　*Plectothyrella platystrophoides* Temple. P. 412, pl. 20, figs. 1 – 5; pl. 21, figs. 1 – 10.
1967　*Plectothyrella platystrophoides* Temple; Marek et Havliček. P. 284, pl. 1, figs. 14 – 17, 19.
1968　*Plectothyrella crassicosta* (Dalman); Bergström. P. 19, pl. 7, figs. 5 – 8.
1974　*Plectothyrella crassicosta* (Dalman); Rong. 199 页，图版 93, 图 29 – 33。
1975　*Plectothyrella platystrophoides* Temple; Fu. 113 页，图 24, 图 6, 7。
1977　*Plectothyrella crassicosta* (Dalman); Zeng. 67 页，图版 23, 图 1, 2。
1979　*Plectothyrella crassicosta* (Dalman); Rong. 2 页，图版 2, 图 8 – 10, 12 – 17。
1981　*Plectothyrella crassicosta* (Dalman); Chang. 563 页，图版 1, 图 37。
1983　*Plectothyrella crassicosta* (Dalman); Zeng. 121 页，图版 15, 图 28 – 32。
1984　*Plectothyrella crassicosta* (Dalman); Rong. P. 163, pl. 14, figs. 5 – 8, fig. 17.
1999　*Plectothyrella crassicosta chauveli* Havliček; Villas, Lorenzo et Marco. P. 193, fig. 4j – o.

2000 *Plectothyrella crassicosta* (Dalman); Carlson et al., P. 1065, fig. 720, 1a – 1e.
2006 *Plectothyrella crassicosta* (Dalman); Rong. 294页,图版2,图8。

描述:贝体中等大,通常壳长9~17mm,壳宽10~20mm(表38);轮廓亚圆形;侧视强双凸型;铰合线短,约为最大壳宽的3/5;背、腹铰合面都小而低(图版48,图4a,4b);主端阔圆;前接合缘轻微单褶型;背中隆低,仅发育在近前缘处;腹中槽浅宽,始于腹壳中后部(图版47,图7a);腹喙较大,弯曲在背喙之上(图版48,图4a,4b)。壳表饰粗圆放射褶,在近前缘处偶尔作1次分叉;腹中槽内4~5根,背中隆上4~5根。

腹内:铰齿粗壮,三角状;齿板中等长,近平行延伸在腹肌痕面两外侧(图版47,图7b;图版48,图4a,4b;插图32-A);腹肌痕面深凹,呈纵长方形或近梯形,其两外侧被齿板所限制。

表38 粗线小褶窗贝介壳测量(单位:mm)
Table 38 Shell measurements of *Plectothyrella crassicosta* (Dalman) (in mm)

| 采集号 | 登记号 | 腹壳(ventral valve) | | 背壳(dorsal valve) | |
(Coll. No.)	(Cat. No.)	长(length)	宽(width)	长(length)	宽(width)
DH3	HB639	17	19		
WH2	HB643				约13
WH3	HB129				约14.5
DH3	HB711	9	约10		
DH3	HB107				约12
WH2	HB642				约10
WH3	HB136				约16
WH2	HB641				约20
FH2	HB748				16
WH3	HB644				约17
WH2	HB683			约5	约4.4

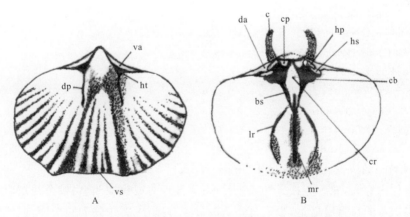

插图32 *Plectothyrella crassicosta* (Dalman)的内部构造
Illustr. 32 Interior structures of *Plectothyrella crassicosta* (Dalman)
A-腹内模(据图版47,图7a,7b);B-背内模(据图版48,图2,6,7b)
A - Ventral internal mold (from pl. 47, figs. 7a, 7b); B - Dorsal internal mold (from pl. 48, figs. 2, 6, 7b)
bs-腕基支板(brachiophore support);c-腕棒(crura);cb-腕棒基(crural base);cp-盖板(cover plate);cr-腕棒腔(crural-ium);da-背铰合面(dorsal area);dp-齿板(dental plate);hp-铰板(hinge plate);hs-铰窝(hinge socket);ht-铰齿(hinge tooth);lr-侧脊(lateral ridge);mr-中隔脊(median ridge);va-腹铰合面(ventral area);vs-腹中槽(ventral sinus)

背内：铰窝长椭圆形,位于铰板前侧；内铰窝脊短粗,有时与强大的腕基融合在一起；铰板分离,呈窄板状,位于铰窝后侧(图版48,图7a,7b；插图32-B)；在2块铰板内侧之间的背窗腔后端具有2个紧靠着的盖板(cover plates),此2个盖板深凹,共同组成如碗状(图版48,图2,7a,7b；图版60,图1,2)(有可能为深凹状的背三角双板)；腕基极为粗壮,其顶端伸出很长的,并从2个盖板外侧方之下,强烈向腹内后方弯曲的腕棒(crura)(图版48,图7b；插图32-B)；腕基支板长,有时相向延伸,并与背中隔脊后部会合成简单腕棒腔(simplex cruralium)(图版48,图7-9),而不是以往文献所说的隔板槽(septalium),但有时腕基支板只延伸在背中隔脊的两侧,不与背中隔脊会合(图版48,图2,6)；背中隔脊长,延伸至背壳底近前缘处；背中隔脊两外侧各具一根向外弯曲的弧形状侧脊；背肌痕面呈月牙状,位于背中隔脊和弧形脊之间(图版48,图2,6,7b-9)。

产地层位：湖北宜昌丁家坡、分乡、王家湾；上奥陶统顶部五峰组观音桥段(赫南特阶中部)。

无窗贝目 Order Athyridida Boucot, Johnson et Staton, 1964
　　无窗贝亚目 Suborder Athyridina Boucot, Johnson et Staton, 1964
　　　小双分贝超科 Superfamily Meristelloidea Waagen, 1883
　　　　小双分贝科 Family Meristellidae Waagen, 1883
　　　　　小双分贝亚科 Subfamily Meristellinae Waagen, 1883

欣德贝属 Genus *Hindella* Davidson, 1882

1882　　*Hindella* Davidson. P. 130.
1951　　*Meristina* Hall; Williams. P. 111.
1965　　*Hindella* Davidson; Boucot, Johnson et Staton. H658.
1967　　*Cryptothyrella* Cooper; Marek et Havliček. P. 283.
1968　　*Cryptothyrella* Cooper; Wright. P. 356.
1974　　*Meristina* Hall; Rong, Xu et Yang. 202页。
1977　　*Meristina* Hall; Zeng. 66页。
1977　　*Hindella* Davidson; Sheehan. P. 34.
1978　　*Meristina* Hall; Yan. 224页。
1979　　*Hindella* Davidson; Rong. 2页。
1981　　*Hindella* Davidson; Chang. 563页。
1981　　*Cryptothyrella* Cooper; Rong et Yang. 243页。
1983　　*Hindella* Davidson; Zeng. 122页。
1984　　*Hindella* Davidson; Rong. 164页。
1987　　*Cryptothyrella* Cooper; Temple. P. 123.
2000　　*Hindella* Davidson; Alvarez, Rong et Boucot. P. 1563.
2006　　*Hindella* Davidson; Rong. 294页。

属型种：Genotype *Athyris umbonata* Billings, 1862.

特征简要：贝体中等大,轮廓亚长方形或亚梨形；侧视近等或腹双凸型；铰合线短而弯曲；背、腹铰合面都很低；腹喙强烈弯曲在背喙之上,腹三角孔全被腹喙遮掩；无背中隆和腹中槽,或偶见于贝体近前缘处,但很微弱。壳表前部具有微弱同心层。

腹内：齿板粗长,中部轻微异向弯曲,限制着深凹的腹肌痕面；腹肌痕面位于两齿板之间,纵长方形,并深凹成深槽状,其前坡较陡,并具数条纵脊；齿板外侧区经常形成椭圆状体腔区,而体腔区内经常显现肋骨状排列的脉管痕。

背内：铰窝呈细长沟状,并向后斜伸在背三角腔两外侧；铰板显著,分离,呈短板状,位于铰窝后侧；内铰窝脊呈脊状；腕基粗壮,位于背三角腔两前侧,被隔板槽所支持；隔板槽深凹,呈窄长"V"字形,是由背中隔板后部分叉而成;背中隔板极长,近伸达背壳底前缘;腕螺强大,螺顶指向贝体侧边。

比较:本属贝体形态和背、腹内部构造特征与 *Meristina* Hall(1867)很相似。它们的主要区别是 *Hindella* 的齿板粗长,限制着深凹、纵长的腹肌痕面,隔板槽窄长,背中隔板极长,伸达至近前缘;而 *Meristina* 的齿板细短,腹肌痕面较短小,隔板槽短,背中隔板仅伸达背壳底中后部。

分布及时代:世界各地;晚奥陶世赫南特期中期至志留纪 Llandoverian 世。

厚欣德贝始端亚种 *Hindella crassa incipiens* (Williams)

图版(pl.)45,图(figs.)6-11;图版(pl.)46,图(figs.)1-10;

图版(pl.)58,图(figs.)1-3;插图(Illustr.)33,A-B

1951	*Meristina crassa incipiens* Williams;P. 112,pl. 6,figs. 14-17;fig. 16a-16c.
1967	*Cryptothyrella* sp. Marek et Havliček. P. 283,pl. 1,figs. 9,10,12,13.
1968	*Cryptothyrella crassa incipiens*(Williams);Wright. P. 356,figs. 3a-3d.
1974	*Meristina crassa incipiens* Williams;Rong,Xu et Yang. 202 页,图版 93,图 27,28。
1977	*Meristina crassa incipiens* Williams;Zeng. 66 页,图版 22,图 7-9。
1977	*Hindella crassa incipiens*(Williams);Sheehan. P. 34,pl. 2,figs. 1-11;pl. 3,figs. 22-24.
1978	*Meristina crassa incipiens* Williams;Yan. 228 页,图版 62,图 12-14。
1979	*Hindella crassa incipiens*(Williams);Rong. 2 页,图版 2,图 7,11。
1981	*Hindella crassa incipiens*(Williams);Chang. 563 页,图版 1,图 38。
1981	*Hindella crassa*(Sowerby);Chang. 564 页,图版 1,图 39。
1983	*Hindella crassa incipiens*(Williams);Zeng. 122 页,图版 15,图 19-22。
1983	*Hindella crassa*(Sowerby);Zeng. 122 页,图版 15,图 23-25。
1984	*Hindella crassa incipiens*(Williams);Rong. P. 164,pl. 14,figs. 1-4.
1987	*Cryptothyrella crassa*(Sowerby);Temple. P. 123,pl. 15,figs. 6,14-25.
2006	*Hindella crassa incipiens*(Williams);Rong. 294 页,图版 2,图 18。

描述:贝体中等大,通常壳长 9.5~17.6mm,壳宽 8.3~20mm(表 39),而且壳宽往往大于壳长,但有时壳长也稍微大于壳宽;轮廓亚圆形或亚梨形;侧视腹双凸型;铰合线短,轻微向后弯曲;主端阔圆;无背中隆和腹中槽。腹壳凸度相对较大,最大凸度位于腹壳中前部;腹喙较大,强烈弯曲在背喙之上,并且掩盖了腹三角孔(图版 46,图 7);腹铰合面小,而且较低。背壳凸度低缓,凸度较为匀称;背铰合面很低;背喙较小,隐藏在腹喙之下。壳表具微弱同心层。

腹内:齿板粗长,其中部轻微异向弯曲在纵长腹肌痕面的两外侧;腹肌痕面纵长,强烈深凹呈深槽状,其前部形成斜坡状,并且有 4~5 条纵脊(图版 46,图 1,2,5,6,9;插图 33-A);齿板两外侧区经常形成椭圆状体腔区,其内经常显露肋骨状排列脉管痕(图版 46,图 1-6)。

背内:铰窝细长沟状,并且向后斜伸在背三角腔两中侧;铰板分离,呈短板状,位于铰窝后侧(图版 46,图 10a,10b;图版 58,图 3;插图 33-B);内铰窝脊呈短脊状,腕基高强,位于背三角腔两前侧,并且被隔板槽两后侧端所支持;隔板槽窄长,由背中隔板后部分叉所形成(图版 46,图 10a,10b;图版 58,图 3;插图 33-B);背中隔板极长,向前伸达背壳底前缘。腕螺形态未显露。

讨论:本书图版 45,图 6-11 等标本,其齿板较短小,腹肌痕面较弱小,隔板槽和背中隔板都不太发育。从这些特征来看,它们似乎更接近于 *Meristina* Hall(1867)的属征。因此那些标本有可能应归于 *Meristina* 的某一个种,这还有待于今后获得更多的标本之后才再进行更进一步的鉴别。

产地层位:湖北宜昌黄花场、丁家坡、王家湾;上奥陶统顶部五峰组观音桥段(赫南特阶中部)。

表 39　厚欣德贝始端亚种介壳测量(单位:mm)

Table 39　Shell measurements of *Hindella crassa incipiens*(Williams)(in mm)

采集号 (Coll. No.)	登记号 (Cat. No.)	腹壳(ventral valve)		背壳(dorsal valve)	
		长(length)	宽(width)	长(length)	宽(width)
WH2	HB663	9.5	8.3		
WH3	HB656	16	20		
WH3	HB650	15	18		
WH3	HB121	14	18.6		
WH2	HB655	12	17		
HK3	IV45659	11.5	13.5		
DH3	HB333			11	
DH2	HB98	11	11.5		
DH3	HB665			5.6	6
WH2	HB654		13		
HK2	IV45655	15.5	14.5		
WH2	HB661	17.6	16.2		
WH1	HB19			14.5	9.2
HK2	IV45741			12	11
WH3	HB709			13	11

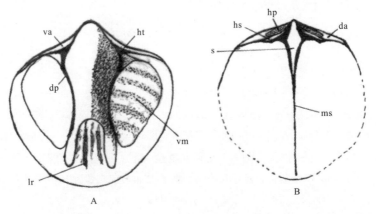

插图 33　*Hindella crassa incipiens*(Williams)的内部构造

Illustr. 33　Interior structures of *Hindella crassa incipiens*(Williams)

A-腹内模(据图版 46,图 2,6);B-背内模(据图版 46,图 10a,10b)

A - Ventral internal mold(from pl. 46,figs. 2,6);B - Dorsal internal mold(from pl. 46,figs. 10a,10b)

da-背铰合面(dorsal area);dp-齿板(dental plate);hp-铰板(hinge plate);hs-铰窝(hinge socket);
ht-铰齿(hinge tooth);lr-纵脊(longitudinal ridges);ms-中隔板(median septum);s-隔板槽(septalium);
va-腹铰合面(ventral area);vm-脉管痕(vascular markings)

优美欣德贝? *Hindella? elegans* Zeng, Zhang et Peng(sp. nov.)

图版(pl.)46,图(fig.)11;图版(pl.)47,图(figs.)1-6;

图版(pl.)59,图(figs.)1-3;插图(Illustr.)34

词源：Elegans(拉丁文),优美,表示新种化石保存美好。

描述：贝体中等大,通常壳长9～16.3mm,壳宽8～16.4mm(表40),轮廓亚圆形或亚梨形;侧视腹双凸型;铰合线短,微弯;主端阔圆状;无背中隆和腹中槽。腹壳凸度相对较强,最大凸度在腹壳中前部,两侧区呈缓坡状,但前坡区坡度较大;腹喙较大,强烈弯曲在背喙之上(图版47,图2);腹三角孔隐藏在腹喙之下;腹铰合面低,轻微弯曲。背壳凸度低于腹壳,整个壳面凸度缓和,而且较匀称;背喙较小,隐伏在腹喙之下;背三角孔也被隐藏;背铰合面窄小。壳表饰有稀同心线。

表40 优美欣德贝? 介壳测量(单位:mm)

Table 40 Shell measurements of *Hindella? elegans* (sp. nov.) (in mm)

采集号 (Coll. No.)	登记号 (Cat. No.)	腹壳(ventral valve)		背壳(dorsal valve)		备注 (remarks)
		长(length)	宽(width)	长(length)	宽(width)	
DH3	HB664			16.3	16.4	
WH2	HB652	15.3	12.6			正型(holotype)
DH3	HB667			9	8.7	
DH2	HB352	9.2	8			
WH2	HB558				13	副型(paratype)

腹内：齿板短而强,相向延伸在腹三角腔两前侧(图版46,图11;图版47,图3,5;插图34-A);腹肌痕面呈近纵长方形,深凹成深沟状,其前坡呈斜坡状,同时具有4～6根纵脊;腹肌痕面两侧被异向弯曲的弧形脊所限制(图版47,图3,5;插图34-A);腹壳底两侧区各形成一个椭圆形状体腔区。

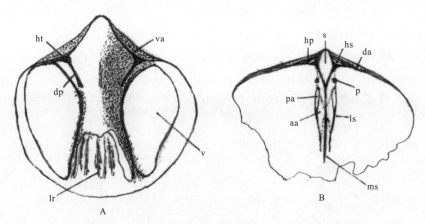

插图34 *Hindella? elegans* (sp. nov.)的内部构造

Illustr. 34 Interior structures of *Hindella? elegans* (sp. nov.)

A-腹内模(据图版47,图3);B-背内模(据图版47,图6;图版59,图3)

A - Ventral internal mold (from pl. 47, fig. 3); B - Dorsal internal mold (from pl. 47, fig. 6; pl. 59, fig. 3)

aa-前闭肌痕(anterior adductor scars);da-背铰合面(dorsal area);dp-齿板(dental plate);hp-铰板(hinge plate);hs-铰窝(hinge socket);ht-铰齿(hinge tooth);lr-纵脊(longitudinal ridges);ls-侧隔板(lateral septum);p-腕锁突起?(jugal process?);pa-后闭肌痕(posterior adductor scars);s-隔板槽(septalium);v-体腔区(visceral area);va-腹铰合面(ventral area)

背内：铰窝呈细沟状，位于铰板内侧；铰板显著，分离，位于细沟状的铰窝外侧（图版 47，图 6；图版 59，图 3；插图 34-B）；内铰窝脊短脊状，向后斜伸；腕基高强，位于背三角腔两前侧；隔板槽窄长，其两后侧端支撑着腕基；背中隔板后部分叉形成隔板槽；背中隔板近伸达至背壳底前部，其前端两侧各具 1 根与背中隔板近平行的细纵脊（图版 47，图 1a, 2, 4, 6）；在背中隔板中部两侧具有 2 对微弱、细长椭圆状的闭肌痕面；在背三角腔两前侧具有 1 对微弱腕锁突起？(jugal process?)（插图 34-B）。腕螺情况不明。

比较：*Hindella? elegans*（sp. nov.）的外形、腹肌痕面的形态，以及主基和隔板槽特征等方面都与 *Hindella crassa incipiens*（Williams, 1951）很相似，它们的主要区别是 *H.? elegans*（sp. nov.）的齿板很短，仅相向延伸在腹三角腔两前侧；纵长、深凹的腹肌痕面两侧则由异向弯曲的弧形脊所限制，细长的背中隔板前部两侧各具 1 根与背中隔板近平行的纵脊；而 *H. crassa incipiens*（Williams）的齿板很粗长，延伸至纵长而深凹的腹肌痕面的两前侧，并限制着腹肌痕面，细长的背中隔板的两前侧无纵脊。

产地层位：湖北宜昌丁家坡、王家湾；上奥陶统顶部五峰组观音桥段（赫南特阶中部）。

6.2 罗惹坪期一些腕足类的新材料

扭月贝目 Order Strophomenida Öpik, 1934
  褶脊贝超科 Superfamily Plectambonitoidea Jones, 1928
    准小薄贝科 Family Leptellinidae Ulrich et Cooper, 1936
     准小薄贝亚科 Subfamily Leptellininae Ulrich et Cooper, 1936
      准小薄贝属 Genus *Leptellina* Ulrich et Cooper, 1936

  属型种：Genotype *Leptellina tennesseensis* Ulrich et Cooper
  小墨西贝亚属 Subgenus *Leptellina* (*Merciella*) Lamont et Gilbert, 1945

1945 *Leptella* (*Merciella*) Lamont et Gilbert, P. 655.
1965 *Merciella* Lamont et Gilbert; Willams. H376.
1981 *Merciella* Lamont et Gilbert; Rong et Yang. 170 页。
2000 *Leptellina* (*Merciella*) Lamont et Gilbert; Cocks et Rong. P. 319.

  亚属型种：*Leptella* (*Merciella*) *versper* Lamont et Gilbert.
  分布及时代：欧洲、中国；志留纪（兰多维列世）。

线纹小墨西贝 *Leptellina* (*Merciella*) *striata* Rong, Xu et Yang
图版 (pl.) 64，图 (figs.) 1-3

1974 *Merciella striata* Rong, Xu et Yang. 198 页，图版 93，图 9, 10。
1977 *Merciella striata* Xu; 曾庆銮, 57 页，图版 19，图 8, 9。
1981 *Merciella striata* Xu, Rong et Yang; 戎嘉余，杨学长, 170 页，图版 2，图 1-16，插图 5。

描述：贝体小，通常壳长 3.2~5.5mm，壳宽 5.3~9mm；轮廓近半圆形；铰合线直，等于最大壳宽；主端锐角状；腹铰合面中等高度，斜倾型；背铰合面低，超倾型。贝体侧视凹凸型；腹壳凸度适度高，顶区凸度较大；背壳浅凹，中部较凹。壳表饰 2 组壳纹，细壳纹较多，夹在较粗壳纹之间。假疹壳。

腹内：齿板较短，仅延伸在腹肌痕面两侧，其前端不会合。腹肌痕面横宽，轻微双叶状；闭肌痕与启肌痕不易区分。

背内：铰窝长三角形；内铰窝脊粗宽，强烈异向展伸；背窗台浅小；主突起单脊状，粗强，其侧边与内铰窝脊后端相融合（图版 64，图 1b）；纤毛环台极为宽大，前端作成双叶状，但台前的围脊较弱；台内的放射脊较弱，仅在台前缘处较显著；背中隔脊短粗，其前端分叉，伸达纤毛环台前缘。

讨论：当前标本的特征以及产出层位都与戎嘉余等（1974, 1981）采集于宜昌大中坝彭家院组（即本书罗惹坪组下段)所描述的 *Merciella striata* 特征相符合。因此当前的标本应归于 *Leptellina*(*Merciella*)*striata* Rong,Xu et Yang(1974)。但以往有些文献将该种的建立者写成 Xu,如曾庆銮(1977,57页);或写成 Xu,Rong et Yang,如戎嘉余等(1981,170页)、曾庆銮(1987,151页),上述这些都是错误的,应予以修正。

产地层位:湖北宜昌杨家湾、王家湾(大中坝);兰多维列统罗惹坪组下段中部。

似小薄贝属 Genus *Leptelloidea* Jones,1928

1928 *Leptelloidea* Jones. P. 388.
1956 *Leptelloidea* Jones;Cooper. P. 763.
2000 *Leptelloidea* Jones;Cocks et Rong. P. 322.

属型种:*Plectambonites leptelloides* Bekker.

分布及时代:欧洲、亚洲;奥陶纪至留纪(兰多维列世)。

志留似小薄贝(新种) *Leptelloidea silurica* Zeng,Zhang et Peng(sp. nov.)
图版(pl.)64,图(figs.)4-7

词源:Silurian(英文),志留纪的,表示本新种的标本是产于志留纪地层。

描述:贝体小,通常壳长 5～8.1mm,壳宽 8～10.5mm(表 41);轮廓近半圆形;侧视强凹凸型;铰合线直,等于最大壳宽;主端近直角状。腹壳凸度中等,最大凸度位于顶区;腹喙较大,微弯;腹铰合面适度高,斜倾型。背壳深凹,中部最低;背喙小;背铰合面较低,超倾型。壳饰有 2 组粗细不等的壳纹,细壳纹较多,夹在粗壳纹之间。假疹壳。

表 41 志留似小薄贝(新种)介壳测量(单位:mm)
Table 41 Shell measurements of *Leptelloidea silurica* (sp. nov.) (in mm)

采集号 (Coll. No.)	登记号 (Cat. No.)	腹壳(ventral valve)		背壳(dorsal valve)		备注 (remarks)
		长(length)	宽(width)	长(length)	宽(width)	
Pm065-6-1F	YB7	6.8	9.2			
标本同上	同上			6.1	9.2	正型(holotype)
Pm065-6-1F	YB4	8.1	10.5			副型(paratype)
Pm065-6-1F	YB9	5	8.0			
Pm065-6-1F	YB73	6.5	9.3			

腹内:齿板发育,相向沿着腹肌痕面延伸,并在腹肌痕面前方中部会合,组成腹肌痕面围脊(图版 64,图 4a,4b);腹肌痕面前部中肌隔短而显著;腹肌痕面双叶状。

背内:铰窝长椭圆状;外铰窝脊高,椭圆脊状;内铰窝脊显著,短脊状(图版 64,图 5b);腕基支板短小、微弱,相向延伸在背窗台两前侧(图版 64,图 5c);主突起强壮,三叶型(图版 64,图 5b);纤毛环台发育,双叶状;背中隔脊高强,前部分叉,并只限在纤毛环台内。

讨论:从当前标本的贝体小,轮廓近半圆形,侧视强烈凹凸型,壳表饰 2 组粗细不同的壳纹;腹肌痕面前缘具围脊,双叶型;背内主突起强烈三叶型;纤毛环台大,双叶状;背中隔脊高强,前部分叉,并只限在纤毛环台内等许多重要特征来看,这些标本应归于 *Leptelloidea* Jones(1928)。据 Cocks and Rong(2000,P.322)的资料,*Leptelloidea* 这个属只产在奥陶纪地层中。然而,当前的标本却产在宜昌地区兰多维列统罗惹坪组下段中部。因此本新种很可能是 *Leptelloidea* 这个属唯一在兰多维列统中发现的一个种。*Leptelloidea silurica*(sp. nov.)与 *L. leptelloides*(Bekker,1921)的主要区别是前者的贝体前缘较阔圆，纤毛环台相对较小，而后者贝体前缘较尖突，纤毛环台很大。

产地层位：湖北宜昌杨家湾；兰多维列统罗惹坪组下段中部。

正形贝目 Order Orthida Schuchert et Cooper,1932
　正形贝亚目 Suborder Orthidina Schuchert et Cooper,1932
　　褶正形贝超科 Superfamily Plectorthoidea Schuchert et LeVene,1929
　　　弓正形贝科 Family Toxorthidae Rong,1984

微小正形贝属(新属) Genus *Minutorthis* Zeng,Chen et Zhang(gen. nov.)

属型种：Genotype *Salopina*? *yichangensis* Rong et Yang,1981.

词源：Minute(英文)，微小的；Orthis 是腕足动物正形贝类的一个属名，表示新属是正形贝类中微小的一种类型。

特征简要：贝体很微小；轮廓近半圆形；侧视平凸型；铰合线直，等于最大壳宽；主端锐角状。腹肌痕面双叶状，无腹中隔板。背内主基很开阔，其宽度大于铰合线长度的 1/3 以上；铰窝小椭圆状；腕基呈宽板状，强烈异向展伸；无腕基支板；主突起单脊状，短粗；背肌痕面不清晰；背中板低弱，位于背壳底中部。壳纹简单，不分枝。无疹壳。

描述：贝体极微小；轮廓近半圆形；侧视平凸型；铰合线直，等于最大壳宽；主端锐角状或近直角状。腹壳凸度中等，在腹喙稍微前方凸度较强，腹喙小；腹铰合面低，斜倾型。背壳平坦；在背喙前方隆起呈小圆丘状；壳表中前部微凹，但不足为背中槽；背三角孔洞开；背铰合面低，正倾型。壳表饰简单放射纹，不分枝；无疹壳质。

腹内：铰齿小，椭圆脊状；齿板短小；腹肌痕面双叶状；闭肌痕居中，呈小长方形状，其前端短于两侧的启肌痕；启肌痕较大，呈半月形，位于闭肌痕两外侧，其前端稍微长于闭肌痕(图版 65，图 6a；插图 35 - A)；无腹中隔板。

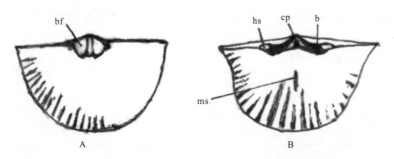

插图 35 *Minutorthis yichangensis*(Rong et Yang)(gen. nov.)的内部构造
Illustr. 35 Interior structures of *Minutorthis yichangensis*(Rong et Yang)(gen. nov.)
A-腹内模(据图版 65，图 6a)；B-背内(据图版 65，图 6b)
A - Internal mold of ventral valve(from pl. 65,fig. 6a)；B - Dorsal interior(from pl. 65,fig. 6b)
b-腕基(brachiophores)；bf-双叶状腹肌痕(biolobed ventral muscle field)；cp-主突起(cardinal process)；hs-铰窝(hinge socketes)；ms-中隔板(median septum)

背内：主基很开阔，其宽度大于铰合线长度的 1/3 以上；铰窝小椭圆状，两者间的间距很大；腕基呈宽板状，强烈异向展伸，其前端轻微向前突伸(图版 65，图 6b；插图 35 - B)；无腕基支板；单脊状主突起；背肌痕面不清晰；背中隔板很低、弱，位于背壳底纵中部。

讨论：当前标本产于宜昌杨家湾剖面罗惹坪组下段中部，与戎嘉余等(1981)采自于宜昌大中坝罗惹坪组的 *Salopina*? *yichangensis* Rong et Yang 可能是同一层位，而且贝体都极其微小，内、外部特征也几乎相同。因此我们认为当前的标本应该是 Salopina? yichangensis Rong et Yang(1981)。但我们将当前保存很好的标本仔细与 Salopina Boucot(1960)进行比较,发现它们彼此之间有着本质的差别,认为应另建立 Minutorthis(gen. nov.)这个新属。Minutorthis(gen. nov.)的贝体极微小(腹壳长 1mm,壳宽 1.3mm);轮廓半圆形;侧视平凸型;铰合线直长,等于最大壳宽;主端锐角状。腹肌痕面双叶状,无腹中隔板。主基极为开阔,其宽度大于最大壳宽 1/3 以上,无腕基支板;主突起单脊状,短粗;背肌痕面不清晰;背中隔板极低弱,仅位于背壳底纵中部。壳纹简单,不分叉。无疹壳。上述这些重要特征与 Salopina 的属征有着本质的差别。

根据 Minutorthis(gen. nov.)的贝体极微小、侧视平凸型、主基极为开阔、铰合线直长、主端锐角状、壳纹简单、无疹壳等特性看,它应归入 Superfamily Plectorthoidea Schuchert et LeVene(1929)当中的 Family Toxorthidae Rong(1984)这个科内,并且与 Toxorthis Temple(1965)的贝体横宽,侧视双凸型,壳线粗而少,具有明显的腹中隆和背中槽的特征可以区分。

另值得一提的是:Minutorthis(gen. nov.)的贝体极微小,又保存着两壳张开,但又铰合在一起,它与 Spinochonetes Rong et Yang 以及 Spinolella Zeng,Zhang et Li(2015)两个属保存在一起,而且该两个属的贝体也很微小,也保存着两壳张开,同时又铰合在一起。表明当时它们是生存在水体很深,又很安宁的一种生态环境中。因此这个腕足动物群应该是底栖组合5(BA5)典型的标志物,从而也表明当时宜昌地区沉积罗惹坪组下段中部时的生态环境应属于底栖组合 4-5(BA4-BA5)的一种外陆架区海域。

分布及时代:中国中南部;兰多维列世晚埃隆期(Late Aeronian)。

宜昌微小正形(新属)Minutorthis yichangensis (Rong et Yang)(gen. nov.)

图版(pl.)65,图(figs.)6a-6b;插图(Illustr.)35

1981 Salopina? yichangensis Rong et Yang,169 页,图版 1,图 28,29;插图 4。

描述:贝体极微小,腹壳长 1mm,腹壳宽 1.3mm,背壳长 0.8mm,背壳宽 1.3mm;轮廓近半圆形;侧视平凸型;铰合线直,等于最大壳宽;主端锐角状或近直角状;腹、背喙部都很小。腹壳凸度中等,在腹喙稍前方的凸度较大;腹铰合面低,斜倾型;腹三角孔情况不明。背壳平坦,在背喙前方隆凸呈小圆丘状;在背壳表中前部轻微凹下,但不足为背中槽;背三角孔洞开;背铰合面低,正倾型。壳表饰简单放射纹,不分叉。无疹壳。

背、腹内构造同属征。

产地层位:湖北宜昌杨家湾;兰多维列统罗惹坪组下段中部。

五房贝目 Order Pentamerida Schuchert et Cooper,1931
  五房贝亚目 Suborder Pentameridina Schuchert et Cooper,1931
    五房贝超科 Superfamily Pentameroidea M'Coy,1844
      五房贝科 Family Pentameridae M'Coy,1844

槽五房贝属 Genus Sulcipentamerus Zeng,1987

1987 Pentamerus(Sulcupentamerus)Zeng. 240 页。
2002 Harpidium(Sulcipentamerus)Zeng;Boucot,Rong et Blodgett. P. 976.
2007 Sulcipentamerus Zeng;Rong,Jin et Zhan. P. 254.

属型种:Genotype Pentamerus(Sulcupentamerus)sulcus Zeng,1987.

分布及时代:中国中南部;晚埃隆期(兰多维列世)。

背平槽五房贝 *Sulcipentamerus dorsoplanus* (Wang)

图版(pl.)65, 图(figs.)1-5; 插图(Illustr.)36

1955 *Pentamerus dorsoplanus* Wang, 130页, 图版69, 图5-8, 15。
1964 *Pentamerus dorsoplanus* Wang; 王钰, 金玉玕, 方大卫, 172页, 图版24, 图7, 8, 10-12。
1977 *Pentamerus dorsoplanus* Wang; 曾庆銮, 55页, 图版18, 图13; 图版19, 图1a-1c。
1978 *Pentamerus dorsoplanus* Wang; 阎国顺, 251页, 图版75, 图1-5。
1981 *Pentamerus dorsoplanus* Wang; 戎嘉余, 杨学长, 208页, 图版11, 图12-21, 24-27; 插图35, 36。
1987 *Pentamerus*(*Sulcupentamerus*)*dorsoplanus* (Wang); 曾庆銮, 242页, 图版18, 图9, 11, 17, 27。
2007 *Sulcipentamerus dorsoplanus* (Wang); Rong, Jin et Zhan. P. 256, pl. 1, figs. 1-6; pl. 2, figs. 5-18; pl. 3, figs. 1-7; pl. 4, figs. 1-11, 16-20。

特征简要: 贝体大, 不作三叶状; 轮廓近长卵形或近亚梨形; 侧视腹双凸型。腹壳凸度较强; 腹喙巨大, 强烈弯曲在背喙之上; 腹三角孔被深凹、呈长菱形腹三角板覆盖(图版65, 图4, 5; 插图36-B)。背壳凸度低缓, 近前缘处具浅凹背中槽(图版65, 图1d)。壳表光滑。

腹内: 腹匙形台窄长, 其长度近为腹壳长度的4/5(图版65, 图2; 插图36-A); 腹中隔板高强, 双板型(图版65, 图4; 插图36-B), 其长度为腹壳长度的2/5。

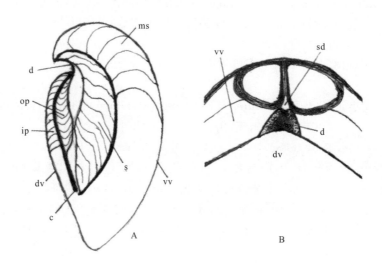

插图36 *Sulcipentamerus dorsoplanus*(Wang)的内部构造

Illustr. 36 Interior structures of *Sulcipentamerus dorsoplanus*(Wang)

A-贝体纵裂面, 示内部构造(据图版65, 图2); B-腹喙部横切面, 示双柱型匙型台和深凹、长菱形腹三角板(据图版65, 图4)

A- Longitudinal section of conjoined valves, showing internal structures(from pl. 65, fig. 2);
B- Transverse section of the ventral beak showing spondylium duplex and a concave deltidium of rhomboidal outline(from pl. 65, fig. 4).

c-腕棒(crura); d-腹三角板(deltidium); dv-背壳(dorsal valve); ip-内铰板(inner hinge plate); ms-中隔板(median septum); op-外铰板(outer hinge plate); s-匙形台(spondylium); sd-双柱型匙形台(spondylium duplex); vv-腹壳(ventral valve)

背内: 内铰板(inner hinge plate)发育, 在背壳底近平行向前延伸, 其长度约近为背壳长度的1/2; 腕棒长, 近伸达至腹匙形台的前端; 外铰板(outer hinge plate)比内铰板稍微宽些, 其前端略微长于内铰板(图版65, 图2, 3; 插图36-A)。

特别说明: 本书中内、外铰板的名称和所在的部位是根据Carlson and al., 2002, P. 927, fig. 621所指定的, 但这与Amsden and al., 1965, H536-537, fig. 406, fig. 407所指定的, 以及被多数学者长期采用的内、外铰板的情况是相反的。

产地层位：湖北宜昌龚家冲；兰多维列统罗惹坪组上段顶部。

中褶贝属（新属）Genus *Centreplicatus* Zeng, Zhang et Han(gen. nov.)

属型种：Genotype *Centreplicatus triangulatus* Zeng, Zhang et Han(gen. et sp. nov.)。

词源：Centre（英文），中心区；Plicate（英文），具褶的，表示新属仅在贝体壳表中心区具有壳褶的一种五房贝类。

特征简要：贝体中等大，不作三叶状；轮廓近三角形；侧视双凸型；前接合缘轻微单槽型或近直缘型。背、腹壳纵中前部各具3根互相对应、低宽的壳褶。内铰板后部近平行向前延伸，但其前端相向内弯，并互相连接成"U"字形。

描述：贝体中等大，不作三叶状；轮廓近三角形；侧视适度双凸型；前接合缘轻微单槽型或近直缘型。腹壳凸度中等，最大凸度位于顶区稍微后部；腹喙大，肿胀，强烈内弯，并紧贴在背喙之上；铰合线短而弯曲；无腹中隆。背壳凸度稍微低于腹壳，在顶区稍微后方的凸度较强；壳表近前方轻微凹下，形成很浅的背中槽，但有的不明显；背喙较小，隐伏在腹喙之下。背、腹壳纵中前部各具3根低、宽，而且是互相对应的壳褶（图版66，图2d），但有的极其微弱；两侧区壳表光滑。

腹内：腹匙形台显著，其长度约为腹壳长的2/3；腹中隔板高强，为双板型，其长度约为腹壳长的1/2（图版66，图4a，4b；插图37-A）。

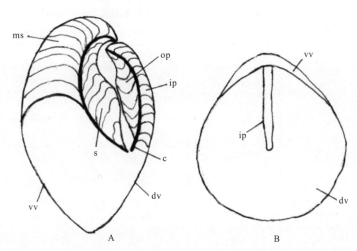

插图37　*Centreplicatus triangulatus*(gen. et sp. nov.)的内部构造

Illustr. 37　Interior structures of *Centreplicatus triangulatus*(gen. et sp. nov.)

A-贝体纵裂面，示内部构造（据图版66，图4a）；B-贝体风化面显示内铰板的形状（据图版67，图5）

A - Longitudinal section of conjoined valves(from pl. 66, fig. 4a); B - Showing the shape of inner hinge plates on weathering front of conjoined valves(from pl. 67, fig. 5)

c-腕棒（crura）；dv-背壳（dorsal valve）；ip-内铰板（inner hinge plate）；ms-中隔板（median septum）；op-外铰板（outer hinge plate）；s-匙形台（spondylium）；vv-腹壳（ventral valve）

背内：内铰板长，沿着背壳底近平行向前延伸，但其前端相向内弯，并连接成"U"字形，其长度约为背壳长的3/5（图版66，图4a，4b；图版67，图5；插图37-A，B）；腕棒长，其前端近伸达至腹匙形台前端；外铰板发育，其宽度大于内铰板，但其前端稍微短于内铰板（图版66，图4a，4b；插图37-A）。

比较：新属*Centreplicatus*(gen. nov.)以其轮廓近三角形，背、腹壳表纵中部各具3根低宽、相对应的壳褶，以及背内的内铰板（inner hinge plate）沿着背壳底近平行向前延伸，但其前端相向内弯，并连接成"U"字形等很独特的特征，明显可和五房贝科（Pentameridae M'Coy, 1844）内的其他各个属相区别。Boucot et al. (1979), pl. 8, figs. 1-13 的 *Harpidium*(*Harpidium*)*kirki* Poulsen 的壳形和壳饰特征与本新属 Centreplicatus(gen. nov.)的外部特征很相似,有可能为本新属的成员,但其内部特征还不了解,有待于今后做进一步的研究去给予证实。

时代及分布:中国中南部;埃隆晚期至特列奇早期(Late Aeronian to Early Telychian)。

三角形中褶贝(新属、新种)*Centreplicatus triangulatus* Zeng, Zhang et Han (gen. et sp. nov.)

图版(pl.)66,图版(figs.)1-4;图版(pl.)67,图版(figs.)1-5;
插图(Illustr.)37-A,B

词源:Triangulatus(拉丁文),三角形的,表示新属、新种的贝体轮廓近三角形。

描述:化石产于湖北宜昌军田坝罗惹坪组上段中部的一层厚17~20cm 的生物灰岩内,化石非常丰富,但非常单调,仅为本新属新种。贝体中等大,不作三叶状;通常壳长20.2~23.6mm,壳宽22.3~26.5mm,壳厚14.9~15.4mm(表42);轮廓近三角形;侧视适度腹双凸型;前接合缘轻微单槽型或近于直缘型。腹壳凸度中等,最大凸度位于顶区稍微后方;腹喙大,肿胀,强烈弯曲,并紧贴在背喙之上;无腹中隆。背壳凸度比腹壳低,在顶区稍微后方的凸度较强;但在背壳表纵中部直至前缘的壳面轻微凹下,成为浅宽背中槽,但有的不明显;背喙短小,隐伏在腹喙之下。壳表纵中部各具3根低宽,而且是两壳相对应的壳褶(图版66,图2d)但有时不清晰;背、腹壳两侧区的壳表光滑,有时见到微弱同心生长线。

表42 三角形中褶贝(新属新种)介壳测量(单位:mm)
Table 42 Shell measurements of *Centreplicatus triangulatus* gen. et sp. nov. (in mm)

采集号 (Coll. No.)	登记号 (Cat. No.)	介壳(shell)			备注 (remarks)
		长(length)	宽(width)	厚(thickness)	
Jl 上 1	HB720	23.4	23.1	15	副型(paratype)
Jl 上 1	HB721	23	23	15	正型(holotype)
Jl 上 1	HB722	21.5	22.1	15.2	
Jl 上 1	HB723	23.6	26.5	15.4	
Jl 上 1	HB724	20.2	22.3	14.9	
Jl 上 1	HB725	16	15.5		风化面(on weathering front)
Jl 上 1	HB726	14	16	10	幼体(junior)

背、腹壳内的构造特征同属征。

产地层位:湖北宜昌军田坝;兰多维列统罗惹坪组上段中部。

从五房贝属 Genus *Apopentamerus* Boucot et Johnson, 1979

1979 *Apopentamerus* Boucot et Johnson, P.104.
1987 *Apopentamerus* Boucot et Johnson; Zeng. 243页。
2002 *Harpidium*(*Isovella*) Breivel et Breivel; Boucot, Rong et Blodgett. P.976.
2007 *Apopentamerus* Boucot et Johnson; Rong, Jin et Zhan. P.260.

属型种:Genotype *Apopentamerus racinensis* Boucot et Johnson, 1979.

分布及时代:北美、中国;晚埃隆期至早罗德洛世(Late Aeronian to Early Ludlowian)。

围板贝亚属（新亚属）Subgenus *Apopentamerus*(*Enclosurus*)
Zeng,Wang et Peng (subgen. nov.)

亚属型种：Subgenotype *Apopentamerus fenxiangensis* Zeng,1987.

词源：Enclosure（英文），围墙、围板之意，表示新亚属整个内脏腔被钙化围板围住的一种很特殊的类型。

特征简述：贝体大，不作三叶状，轮廓亚梨形，侧视双凸型；前接合缘直缘型。腹匙形台、腹中隔板，以及内、外铰板和腕棒的特征犹如 *Apopentamerus*，其最为特殊的是整个内脏腔被钙化围板（calcific enclosures）围住。

描述：贝体大，不作三叶状，轮廓亚梨形，侧视较强双凸型，前接合缘直缘型；铰合线很短，弯曲；腹喙大，肿胀，强烈弯曲，并且高悬在背喙之上（图版68，图1c）。腹壳凸度中等，最大凸度位于顶区，无槽、隆；腹三角孔被深凹、长菱形腹三角板覆盖（图版68，图1a,3a；插图38-Ba）。背壳凸度适当，顶区凸度较大，无槽、隆。壳表光滑。

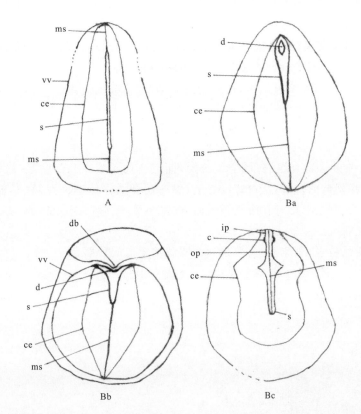

插图38 *Apopentamerus* (*Enclosurus*) *fenxiangensis* (Zeng)(Subgen. nov.)的内部构造

Illustr. 38 Showing interior structures of *Apopentamerus* (*Enclosurus*) *fenxiangensis* (Zeng)(Subgen. nov.)

A-贝体侧剖面，示匙形台、中隔板和钙化围板（据图版68，图2a）；Ba-c：成年贝体的连续横切面，示内部构造（据图版68，图3a-c）

A - Lateral section of conjoined valves,showing spondylium, median septum and calcific enclosures in the visceral area(from pl. 68,fig. 2a)

Ba - Bc - Serial sections of a mature shell showing interior structures(from pl. 68,figs. 3a - 3c).

c-腕棒基(crura base);ce-钙化围板(calcific enclosures);d-腹三角板(deltidium);db-背喙(dorsal beak);ip-内铰板(inner hinge plate);ms-中隔板(median septum);op-外铰板(outer hinge plate);s-匙形台(spondylium);vv-腹壳(ventral valve)

腹内：腹匙形台窄长，其长度大于腹壳长度的2/3（图版68，图2a,2b；插图38-A）。腹中隔板高强，近伸达腹壳底 1/2 处;钙化围板从铰齿下方伸出,然后沿着腹内脏腔周边延伸,并在腹内脏腔前方会合,包围着整个腹内脏腔(图版 68,图 2,3;插图 38-A,B)。

背内:内铰板(inner hinge plates)沿着背壳底近平行向前延伸至背壳底中前部;腕棒基显著;腕棒长,其前端伸达至腹匙形台前端附近;外铰板(outer hinge plates)比内铰板宽,并与腹匙形台张开的宽度作相适应地张开(图版 68,图 3c;插图 38-Bc),其长度与内铰板近等长;在铰窝下方各伸出钙化板,然后沿着背内脏腔周边延伸,包围着整个背内脏腔,并与腹内的钙化板联合,共同包围着整个贝体的内脏腔((图版 68,图 3c;插图 38-Bc)。

讨论:当前标本的贝体内存在着很有规律的钙化围板。腹内的钙化围板是从铰齿下方伸出,而背内的钙化围板则从铰窝下方伸出,并分别沿着内脏腔周边延伸,然后互相联合共同包围着整个内脏腔。钙化围板虽然不是贝体内真正的骨骼,但它们必定是由其贝体内的一种组织结构为钙化提供基础。因此建议把它们当作副骨骼看待,并作为亚属级的分类依据。因此,我们也将 Apopentamerus fenxiangensis Zeng(1987)作为亚属型种,建立 Apopentamerus(Enclosurus)(Subgen. nov.)这个新亚属。

分布及时代:中国中南部;晚埃隆期至早特列奇期(Late Aeronian to Early Telychian)。

分乡围板贝(新亚属)Apopentamerus(Enclosurus)fenxiangensis
(Zeng)(subgen. nov.)

图版(pl.)68,图(figs.)1-3;插图(Illustr.)38-A,B

1987 *Apopentamerus fenxiangensis* Zeng,243 页,图版 16,图 23-27。
2007 *Sulcipentamerus dorsoplanus*(Wang);Rong,Jin et Zhan. P. 256,pl. 2,figs. 1-4.

特征简述:贝体大,壳长 52.2～57mm,壳宽 40～42.8mm,壳厚 33～34mm,不作三叶状;轮廓亚梨形;侧视较强双凸型;前接合缘直缘型;铰合线很短,而且弯曲。腹壳凸度中等,最大凸度位于顶区,无槽、隆;腹喙大,肿胀,强烈弯曲,并高悬在背喙之上;腹三角孔被深凹、长菱形腹三角板覆盖。背壳凸度适当,顶区凸度较大;背喙较小,微弯在腹三角板前方之上;无槽、隆。壳表光滑。

背、腹内构造同亚属征。

产地层位:湖北宜昌分乡、龚家冲;兰多维列统罗惹坪组上段顶部。

参考文献

王传尚,汪啸风,陈孝红,等.奥陶纪末期层序地层研究[J].地球科学——中国地质大学学报,2003,28(1):6-10.
王钰,金玉玕,方大卫.中国的腕足动物化石(上、下册),中国各门类化石[M].北京:科学出版社,1964:1-777.
王钰,金玉玕,方大卫.腕足动物化石[M].北京:科学出版社,1966:1-702.
王淑敏.湖北腕足类[M].//湖北省区域地质测量队,湖北省古生物图册.武汉:湖北科学技术出版社,1984:128-236,图版62-95.
王成源,陈立德,王怿,等.Pterospathodus eopennatus(牙形刺)带的确认与志留系纱帽组的时代及相关地层的对比[J].古生物学报,2010,49(1):10-28.
王英华.碳酸盐岩导论[M].长沙:湖南省地质局,1979:1-234.
中南地区区域地层表编写小组.中南地区区域地层表[M].北京:地质出版社,1974:1-534.
中国科学院南京地质古生物研究所.西南地区地层古生物手册[M].北京:科学出版社,1974:1-454,图版1-202.
戎嘉余.中国的赫南特贝动物群(Hirnantia fauna)并论奥陶系与志留系的分界[J].地层学杂志,1979,3(1):1-27,图版1-2.
戎嘉余.上扬子区晚奥陶世海退的生态地层证据与冰川活动影响[J].地层学杂志,1984,8(1):19-28.
戎嘉余.生态地层学的基础——群落生态的研究[C].//在中国古生物学会第十三、十四届学术年会论文选集.合肥:安徽科学技术出版,1986:1-24.
戎嘉余,许汉奎,杨学长.西南地区志留纪腕足类[M].//在中国科学院南京地质古生物研究所.西南地区地层古生物手册.北京:科学出版社,1974:195-208,图版92-96.
戎嘉余,马科斯·约翰逊,杨学长.上扬子区早志留世(兰多维列世)的海平面变化[J].古生物学报,1984,23(6):672-693.
戎嘉余,陈旭,王成源,等.论华南志留系对比的若干问题[J].地层学杂志,1990,14(3):161-177.
戎嘉余,詹仁斌.赫南特阶腕足类[J].//陈旭.奥陶系上统赫南特阶全球层型剖面和点位的建立.地层学杂志,2006,30(4):289-305,图版2.
戎嘉余,王怿,詹仁斌,等.论桐梓上升志留纪埃隆晚期黔中古陆北扩的证据[J].地层学杂志,2012,36(4):679-691.
汪啸风,倪世钊,曾庆銮,等.长江三峡地区生物地层学(2)[M].//早古生代分册.北京:地质出版社,1987:1-641,图版1-72.
汪啸风,陈孝红,张仁杰,等.长江三峡地区珍贵地质遗迹保护和太古宙—中生代多重地层划分与海平面升降变化[M].北京:地质出版社,2002:1-341.
许庆建,等.四川早古生代腕足类[M].//西南地区古生物图册,四川分册(一).北京:地质出版社,1978:284-381,图版115-151.
陈旭.英国奥陶纪和志留纪笔石带的最新划分方案[J].地层学杂志,2010,34(2):161-164.
陈旭,丘金玉.宜昌奥陶纪的古环境演变[J].地层学杂志,1986,10(1):1-15,图版1-2.
陈旭,戎嘉余.中国扬子区兰多维列统特列奇阶及其与英国的对比[M].北京:科学出版社,1996:1-162.
陈旭,戎嘉余,樊隽轩,等.奥陶系上统赫南特阶全球层型剖面和点位的建立[J].地层学杂志,2006,30(4):289-30,图版1-2.
郑宁,宋天锐,李廷栋,等.华南造山带下寒武统和中奥陶统发现放射虫[J].中国地质,2012,39(1):260-265.
杨学长,戎嘉余.川黔湘鄂边界志留系秀山组上段的腕足类化石群[J].古生物学报,1982,21(4):417-432,图版1-3.
耿良玉,蔡习尧.扬子区志留纪兰多维列统胞石序列[J].古生物学报,1988,27(2):249-255,图版1.
徐桂荣,王永标,龚淑云,等.生物与环境的协同进化[M].武汉:中国地质大学出版社,2005:1-293.
阎国顺.长江三峡地区奥陶纪—志留纪腕足类[M].//湖北省地质局三峡地层研究组.峡东地区震旦纪至二叠纪地层古

生物.北京:地质出版社,1978:1-381,图版1-113.

倪寓南.湖北宜昌早志留世笔石[J].古生物学报,1978,17(4):387-416,图版1-4.

常美丽.湖北宜昌晚奥陶世末期赫南特动物群[J].古生物学报,1981,20(6):557-566,图版1.

傅力浦.西北地区早古生代腕足类[M].//地质矿产部西安地质矿产研究所.西北地区古生物图册,陕、甘、宁分册(一).北京:地质出版社,1982:95-179,图版30-45.

曾庆銮.中南地区早古生代腕足类[M].//湖北省地质科学研究所等.中南地区古生物图册(一).北京:地质出版社,1977:27-69,图版10-23.

曾庆銮.长江三峡地区早古生代腕足类[M].//汪啸风等.长江三峡地区生物地层学(2),早古生代分册.北京:地质出版社,1987:209-245,图版8-18.

曾庆銮,赖才根,徐光洪,等.长江三峡地区奥陶纪[M].//汪啸风等.长江三峡生物地层学(2),早古生代分册.北京:地质出版社,1987:43-142.

曾庆銮,刘印环,王建平,等.东秦岭南部奥陶—志留系界线附近腕足动物群演替及生态[J].古生物学报,1993,32(3):372-383,图版1-3.

曾庆銮,胡昌铭.江西玉山王家坝早志留世早期(Early Llandoverian)新腕足动物群的发现及其意义[J].古生物学报,1997,36(1):1-17,图版1-3.

曾庆銮,张森,李志宏.宜昌地区早志留世(Llandoverian)腕足类一新科Spinochonetidae(fam. nov.)及其意义[J].地质学报,2015,89(4):681-691,图版1-2.

曾庆銮,王传尚,李志宏,等.鄂西地区奥陶纪至早志留世生态地层探究[J].地层学杂志,2015,39(3):329-344.

鲜思远.贵州地区早古代腕足类[M].//西南地区古生物图册,贵州分册(一).北京:地质出版社,1978:251-336,图版94-128.

葛治洲,戎嘉余,杨学长,等.西南地区志留系[M].//西南地区碳酸盐岩生物地层.北京:科学出版社,1979:155-220.

湖北省地质局三峡地层研究组.峡东地区震旦纪至二叠纪地层古生物[M].北京:地质出版社,1978:1-381,图版1-113.

詹仁斌,戎嘉余.江西玉山下镇晚奥陶世扭贝族一新属——*Tashanomena*[J].古生物学报,1994,33(4):416-428,图版1.

詹仁斌,戎嘉余.浙赣边区晚奥陶世腕足动物四新属[J].古生物学报,1995,34(5):549-574,图版1-4.

Alvarez F, Rong J Y. Athyridida[J]. Treatise on Invertebrate Paleontology, Part H, Brachiopoda (Revised), 2002, 4: 1475-1601, figs. 1001-1092.

Amsden T W. Pentameridina[J]. Treatise on Invertebrate Paleontology, Part H, Brachiopoda 1965, 2: H536-552, figs. 406-418.

Amsden T W. Late Ordovician and Early Silurian articulate brachiopods from Oklahoma, Southwestern Illinois, and Eastern Missouri[J]. Oklahoma Geological Survey Bulletin, 1974, 119: 1-154, pls. 1-28.

Bassett M G. The articulate brachiopods from the Wenlocek Series of the Welsh Borderland and south Wales. part 4[J]. Monograph of the Palaeontographical Society, 1977, 130(547): 123-176, pls. 33-47.

Bassett M G. Craniida[J]. Treatise on Invertebrate Paleontology, Part H, Brachiopoda (Revised), 2000, 2: 169-183, figs. 93-107.

Bergström J. Upper Ordovician brachiopods from Västergötland, Sweden[J]. Geologica et Palaeontologica, 1968, 2: 1-35, pls. 1-7.

Boucot A J. Early Paleozoic Brachiopods of the Moose River Synclinorium, Maine[J]. Professional Paper U. S. Geological Survey, 1973, 784: 1-81, pls. 1-23.

Boucot A J. Evolution and Extinction Rate controls[M]. Amsterdam: Elsevier, 1975: 1-427.

Boucot A J, Johnson J G. Pentamerinae (Silurian Brachiopoda)[J]. Palaeontographica (Abt. A. Band) 1979, 163: 87-129, pls. 1-15.

Boucot A J, Rong Jiayu, Blodgett R B. Pentameridina[J]. Treatise on Invertebrate Paleontology, Part H, Brachiopoda (Revised), 2002, 4: 960-1026, figs. 643-697.

Cocks L R M. Some Strophomenacean brachiopods from the British Lower Silurian[J]. Bull Brit Mus (Natural History) Geol, 1968, 15(6): 285-324, pls. 1-14.

Cocks L R M. Silurian brachiopods of the Superfamiy Plectambonitacea[J]. Brit Mus (Natural History), Bulletin (Geology),1970,19(4):139 - 203,pls. 1 - 17.

Cocks L R M, Rong J Y. Classification and review of the Brachiopod superfamily Plectambonitacea[J]. Brit Mus(Natural History),Bulletin(Geology)1989,45(1):77 - 163,figs. 1 - 174.

Cocks L R M, Rong J Y. Strophomenida[J]. Treatise on Invertebrate Paleontology,Part H,Brachiopoda(Revised),2000, 2:216 - 362,figs. 132 - 236.

Cocks L R M,Fortey R A. A new *Hirnantia* fauna from Thailand and the biogeography of the latest Ordovician of south - east Asia[J]. Geobios,1997(97):117 - 126,pls. 1 - 2.

Cooper G A. Chazyan and related brachiopods[J]. Smithsonian Miscellaneous Collections,1956,127(I - II):1 - 125,pls. 1 -269.

Haper D A T. The stratigraphy and fauna of the Upper Ordovician High Mains Formation of the Girvan district[J]. Scott J Geol,1981,17(4):247 - 255,pl. 1.

Haper D A T. Dalmanellidina[J]. Treatise on Invertabrate Paleontology,Part H,Brachiopoda(Revised),2000,3:782 - 844,figs. 566 - 615.

Havliček V. Ramenonozci Ceského Ordoviku(The Ordovician Brachiopoda from Bohemia)[J]. Rozpravy Ústředniho ústavu geol. 13:English translation on,1950: 75 - 135,pls. 1 - 13.

Havliček V. Brachiopoda of the Strophomenidina in Czechoslovakia[J]. Rozpravy Ústředního Ústavu geologického,1967, 33:1 - 235,pls. 1 - 52.

Havliček V. Brachiopods of the order Orthida in Czechoslovakia[J]. Rozpr. Ústr. Úst. Geol. ,Svazek 1977,44:1 - 327,pls. 1 - 56.

Havliček V. Mediterranean and Malvinokaffric Provinces:new date on the Upper Ordovician and Lower Silurian brachiopods[J].Časopis pro mineralogii a geologii,roč,1990,35,č. 1:1 - 14,pls. 1 - 4.

Havliček V,Leonardo Branisa. Ordovician Brachiopods of Bolivia(Succession of assemblages,Climate control,affinity to Anglo - French and Bohemian provinces)[J]. Rozpravy Ceskoslovenské Akademie věd,Praha,1980,90(1):1 - 54,pls. 1 - 4.

Havliček V,Jiři Křiž,Enrico Serpagli. Upper Ordovician brachiopod assemblages of the Carnic Alps,Middle Carinthia and Sardinia[J]. Bullettino della Società Paleontologica Italiana,1986,25(3):277 - 311,pls. 1 - 9.

Holmer L E,Popov L E. Lingulida[J]. Treatise on Invertebrate Paleontology,Part H,Brachiopoda(Revised)2000, 2:30 - 97,figs. 7 - 48.

Lesperance P J,et Sheehan P M. Brachiopoda from the Hirnantian Stage(Ordovician - Silurian)at Perce,Quebec[J]. Palaeont, 1976,19:719 - 731.

Lindström M. On the lower Chasmops Beds in the Fågelsång district (Scania) [J]. Geol. Föreningens i Stockholm Förhandlingar,1953,75(2):125 - 148,pl. 1.

Marek L,et Havliček V. The articulate Brachiopods of the Kosove Formation(Upper Ashgillian)[J]. Vest Ústr Úst Geol, 1967,42:275 - 284,pls. 1 - 4.

Mitchell W I. The Ordovician brachiopods from Pomeroy,Co. Tyrone[J]. Palaeontographical Society Monograph,1977, 130(545):1 - 138,pls. 1 - 28.

Nikitin I F. Brachiopods[J]. Apollonov M K(eds):The Ordovician - Silurian boundary in Kazakhstan. *Nauka* Kazakh. SSR. Pupl. House,Alma - Ata,1980:35 - 74,pls. 10 - 21.

Öpik A A. Lower Silurian fossils from the Ilaenus Band,Heathcote,Victoria[J]. Memoirs of the Victoria Geol Survey, 1953(19):1 - 42,pls. 1 - 13.

Popov L E, Holmer L E. Craniopsida[J]. Treatise on Invertebrate Paleontology,Part H,Brachiopoda(Revised),2000,2: 164 - 168,figs. 89 - 92.

Rong J Y. Brachiopods of latest Ordovician in the Yichang District,Western Hubei,Central China. In Nanjing Institute of Geology and Palaeontology,Academia Sinica compiled. Stratigraphy and palaeontology of systemic boundaries in China [J]. Ordovician - Silurian boundary. Hefei:Anhui Science and Technology Publishing House,1984:111 - 176,pls. 1 - 14.

Rong J Y, Jin J S, Zhan R B. Two new genera of Early Silurian Stricklandioid Branchiopods from South China and their bearing on Stricklandioid classification and palaeobiogeography[J]. Journal of Paleontology, 2005, 79(6):1143-1156, figs. 6-10.

Rong J Y, Jin J S, Zhan R B. Early Silurian Sulcipentamerus and related Pentamerid Brachiopods from South China[J]. Palaeontology, 2007, 50(1):245-266, pls. 5.

Savage N M. Rhynchotrematoidea[J]. Treatise on Invertebrate Paleontology, Part H, Brachiopoda(Revised), 2002, 4: 1047-1091, figs. 707-738.

Sutcliffe O E, Harper D A T, Salem A A, et al. The development of an atypical *Hirnantia* - brachiopod Fauna and the onset of glaciation in the late Ordovician of Gondwana[J]. Earth and Environmental Science Transactions of the Royal Society of Edinburgh, 2001, 92:1-14.

Temple J T. Upper Ordovician brachiopods from Poland and Britain[J]. Acta Palaeontologica Polonica, 1965, 10(3):379-427, pls. 1-21.

Temple J T. The Lower Llandovery(Silurian)Brachiopods from Keisley, Westmorland[J]. Palaeontogr. Soc. (Monogr.), 1968, 122(521):1-58.

Temple J T. The Lower Llandovery brachiopods and trilobites from Ffridd Mathrafal, near Meifod, Montgomeryshire[J]. Palaeontographical Society Monograph, 1970:1-76, pls. 1-19.

Temple J T. Early Llandovery brachiopods of Wales[J]. Palaeontographical Society Monograph, 1987:1-137, pls. 1-15.

Villas E, Lorenzo S, Gutiérrez-Marco J C. First record of a *Hirnantia* Fauna from Spain, and its contribution to the Late Ordovician Palaeogeography of northern Gondwana[J]. Transactions of the Royal Society of Edinburgh: Earth Sciences, 1999, 89:187-197, fig. 4.

Williams A. On Llandovery brachiopods from Wales with special reference to the Llandovery District[J]. Quarterly Journal of the Geological Society of London, 1951, 107(1):85-136, pls. 3-8.

Williams A. The Barr and Lower Ardmillan Series(Caradoc) of the Girvan District, South-West Ayrshire[J]. Geological Society of London, 1962(3):1-267, pls. 6-25.

Williams A. The Caradocian brachiopod fauna of the Bala District, Merionethshire[J]. British Museum(Natural History), Bulletin(Geology), 1963, 8(7):327-471, pls. 1-16.

Williams A. Suborder Strophomenidina[J]. Moore, R. C. (Directed and Edited): Treatise on Invertebrate Paleontology, Part H, Brachiopoda, 1965, 1: H362-412, figs. 231-271.

Williams A. Ordovician Brachiopoda from the Shelve District, Shropshire[J]. British Museum(Natural History), Bulletin (Geology Supplement), 1974, 11:1-163, pls. 1-28.

Williams A, Brunton H C. Orthotetidina[J]. Treatise on Invertebrate Paleontology, Part H, Brachiopoda(Revised), 2000, 3:644-681, figs. 465-492.

Williams A, Harper D A T. Orthida[J]. Treatise on Invertebrate Paleontology, Part H, Brachiopoda(Revised), 2000, 3: 714-782, figs. 516-565.

Wright A D. Triplesiidina[J]. Treatise on Invertebrate Paleontology, Part H, Brachiopoda(Revised), 2000, 3:681-689, Figs. 493-497.

Ziegler A M. Silurian marine communities and their environmental significance[J]. Nature, 1965, 207:270-272.

Abstract

Hirnantia fauna, its habitat and extinctive cause of Middle Hirnatian and biota evolution of Llandoverian in Yichang District, Western Hubei, Central China

Zeng Qingluan[1], Chen Xiaohong[2], Wang Chuanshang[2],
Zhang Miao[2], Han Huiqing[1] and Peng Zhongqin[2]

1. Yichang base of Wuhan Center of China Geological Survey, Yichang 443005, Hubei, China
2. Wuhan Center of China Geological Survey, Wuhan 430223, Hubei, China

1. General situation of the *Hirnantia* Fauna

A very rich and very graceful brachiopods of the *Hirnantia* Fauna were collected from the Guanyinqiao Member(Bed) of the Wufeng Formation in Wangjiawan, Dingjiapo and Huanghuachang sections, respectively, in the northern part of the Yichang District. Guanyinqiao member is 0.17~0.3m thick and is lithologically divided into three layers, namely, the blackish-grey, yellowish-brown calcareous hydromica clay rocks(H1) at the lower layer; the yellowish-grey, cream-colored or light purplish-grey hydromica clay rocks with quartz(or rhyolitic tuffite?)(H2) at the middle layer; and the yellowish-grey or light grey hydromica clay rocks(H3) at the upper layer. They are in conformable contacts with both the overlying blackish-grey carbonaceous mudstone(shale) containing *persculptus* Zone in the bottom of the Longmaxi Formation and the underling blackish-grey or brownish-yellow siliceous hydromica clay rocks yielding *extraordina-rius* Zone(see text illustr. 1,2).

There are 43 species in 35 genera which are described in the Guanyinqiao Member, including 11 new species, 6 new generas(*Minutomena*, *Yichangomena*, *Sinomena*, *Hubeinomena*, *Trimena* and *Paramirorthis*) and 2 new families(Sinomenidae and Dysprosorthidae). The Guanyinqiao Member is characterized by great abundance of brachiopoda of the *Hirnantia* Fauna which is represented by an association of distinctive elements, such as *Hirnantia*, *Kinnella*, *Trucizetina*, *Draborthis*, *Dysprosorthis*, *Mirorthis*, *Leptaenopoma*, *Aphanomena*, *Eostropheodonta*, *Paromalomena* and *Plyctothyrella*. This *Hirnatia* Fauna is usually associated with a few trilobite, i.e., the *Dalmanitina*, *Diacanthaspis*, *Leonaspis* etc., in the lower part of the Guanyinqiao Member.

2. Habitat of the *Hirnantia* Fauna

From the Late Katian to the early Middle Hirnantian, as an expansion result of the ice sheet onto the continental shelf of the Gondwana, the global sealevel fell gradually(regression)(Sutcliffe et al., 2001). A gradual sealevel fall(regression) can also be observed in the Upper Yangtze Sea(see illustr. 4-A,B,C). This event provides an excellent ecologic environment for brachiopods of the *Hirnantia* Fauna in the Middle Hirnantian. The fauna, with extremely large abundance and diversity, is characterized by small or very small shells(see text pls. 1-60) for most of its elements in Yichang District. The corals *Amsassia* sp., *Protoheliolites* sp. yielding in the Guanyinqiao Member can also be found in northern Guizhou and southern Sichuan. Therefore, it is inferred that the *Hirnantia* Fauna may live in lukewarm-water rather than cold-water, in the position of Benthic Assemblage 4(BA4)(see text illustr. 4-C)in Yichang District.

3. Extinctive cause of the *Hirnantia* Fauna

Owing to a stepwise removal of the ice sheet from the continental shelf of Gondwana, a gradual sea-level rise(transgression)(Sutcliffe et al.,2001) occurred in the initial stage of the early Late Hirnantian(the base of the *persculptus* Zone) to late Rhuddanian(Early Silurian). This event can also be observed in the Upper Yangtze Sea (see text illustr. 4-D,E;illustr. 5-A). The mass pelagic graptolite fauna lived in the surface-water of the ocean during that time. Sunlight was kept out and oxygen in the surface-water is inevitably exhausted by the pelagic graptolite fauna, which resulted in a serious lack of oxygen(anoxic water)in the bottom-water of the ocean. Therefore, the anoxic event is the main cause for mass extinction event of the *Hirnantia* Fauna.

4. Biotic evolution of Llandoverian

4.1 The exclusive territory of the pelagic graptolite fauna

In the late Hirnantian and throughout the Rhuddanian(early Silurian), the rapid removal of the ice sheet from the continental shelf is resulted in a gradual global eustatic rise(transgression)(Sutcliffe et al.,2001), which can also be observed in the Upper Yangtze Sea(see text illustr. 4-D,E;illustr. 5-A). A pelagic graptolite fauna, with high abundance and diversity, lived in the Upper Yangtze sea in Yichang District. The graptolite fauna, which is found in the Black shale Member of the Longmaxi Formation, is represented by *persculptus*,*ascensus*,*acuminatus*,*atavus*,*acinaces*,*cyphus* and *triangulatus* etc. Therefore,the Yichang District is an exclusive territory of the pelagic graptolite fauna in that period. The fossils and the lithologic characteristics show strong evidences for a deep-water(open ocean) environment which is located at the position of the Benthic Assemblage 6(BA6),corresponding to the study from Pelagic Graptolite Community of Ziegler(1965)and the Pelagic Community of Boucot (1975).

4.2 Large decline of the Pelagic Graptolite Fauna

As a result of the Tongzi uplift, a gradually eustatic fall(regression) in Yichang District (see text illustr. 5-B,C) occurred in the Middle to Late Aeronian in northern Guizhou and southern Sichuan (Rong et al.,2012). The Yellowish-Green Shale Member of the Longmaxi Formation consists of rapid sediment of sandy siltstone with few fossils, the total thickness is up to 571m. The pelagic graptolite fauna showed a stepwise mass of decline in both abundance and diversity during this regression event. In addition,the "*Lingula*"sp., a typical representative of Benthic Assemblage 1(BA1)(Ziegler, 1965;Boucot,1975;Rong,1986) can be found in the upper part of the Yellowish-Green Shale Member (middle *arcuata* zone). Therefore, the evidences from lithostratigraphy, biostratigraphy, ecostratigrapy and rapid sedimentary facies, support the regressive event in the Upper Yangtze Sea during middle-late Aeronian(Middle Llandoverian) in Yichang District.

4.3 The Great recovery of biota

The evolution of brachiopod communities from the top part of the Yellowish-Green Shale Member of the Longmaxi Formation to the lower part of the upper member of the Luoreping(Lojoping) Formation in Yichang District,reveals the ecologic niche is changing from"*Lingula*"Community(BA1)

→*Nucleospira* Com. (BA2)→*Meifodia* Com. (BA3)→*Spinochonetes* – *Spinolella* Com. (BA4 – 5). This evidence indicated a gradual eustatic rise(transgression) in the Upper Yangtze Sea(see text illustr. 5 – D,E). In addition,the Luoreping Formation contains extremely abundant fossils of the corals, which indicates a warm – water environment. The environment, containing rich oxygen and calcium carbonates, provides a good habitat for the benthonic fauna. Therefore,a great recovery of the biota occurred in the Yichang District during that time. The biota contains 10 taxa, i. e. , the brachiopods, corals,trilobites,conodonts,graptolites,caphalopods,gastropods,crinoids,chitinozoans and bryozoans,of which the brachiopods(28 genera) and corals(19 genera)are the dominant taxa with very high abundance.

4.4 The Upper Yangtze Sea towards doom

In Yichang District,the evolution sequence of brachiopod communities from the upper member of the Luoreping Formation to the top part of the fourth member(the highest member)of the Shamao Formation in ascending order is as follows: *Sinokulumbella*(original *Kulumbella* or *Stricklandia*) Community(BA4)→*Sulcipentamerus*(or. *Pentamerus*)Com. (BA3)→*Katastrophomena* Com. (BA2) →*Nucleospira* Com. (BA1 – 2)→*Nalivkinia* Com. (BA1),which showed a gradul sea – level fall(regression)of the Upper Yangtze Sea(see text illustr. 5 – E,F). The typical sedimentary characteristics of the Shamao Formation are rich in fine – grained and medium – grained sandstone, and rarity of fossils. The total thickness of the Shamao Formation is up to 670.5 m due to the rapid sedimentation. There are many sedimentary structures, i. e. , the bed ripples, cross beddings, rain prints, mud – cracks,etc. The evidence reveals a gradual eustatic fall(regression) of the Upper Yangtze Sea(see text illustr. 5 – F) to a shallow water environment. The parallel unconformity, between the top of the Shamao Formation of the Lower Silurian and the base of the Yuntaiguan Formation of the Middle Devonian, reveals that the cause of the closure of the Upper Yangtze Sea, which continuously lasted about 272 Myr±, is the Epeirogenic Movements(namely the Yangtze Uplift,Rong et al. ,2012).

5. Diagnosis of new families and genera

5.1 Some new families and genera of the *Hirnantia* Fauna

Superfamily Strophomenoidea King,1846
  Family Strophomenidae King,1846
    Subfamily Furcitellinae Williams,1965

           Genus *Minutomena* Zeng,Zhang et Han(gen. nov.)

Genotype:*Minutomena yichangensis* Zeng,Zhang et Han(gen. et sp. nov.)
Diagnosis:Shells are small,subcircular in outline;gently biconvex to very gently resupinate in lateral profile;hinge – line straight,slightly narrower than the greatest width of the shell;the pseudodeltidium and chilidium plates are prominent. The ornamentation consists of unequal, round, radiating costae and a few concentric lines. Shell substance is pseudopunctate.

Its ventral interior has sturdy, triangular teeth;incurvate dental plates,extending to the anterior lateral part to bound transversly wide ventral muscle field.

Its dorsal interior is characterized by a very short cardinalia;large,triangular sockets;strongly and widely divergent inner socket ridges;very short,sturdy bilobed cardinal process;at the dorsal shell bottom,there bears a pair of sturdy,divergent lateral ridge;no dorsal median septum.

Comparison: This new genus is similar to *Furcitella* Cooper(1956), but it differs from the latter in less of dorsal median septum, whereas the *Furcitella* has a strongly bifurcate median ridge.

Range and distribution: From Late Ordovician to Llandoverian; Central China.

Family Sinomenidae Zeng, Chen et Zhang(fam. nov.)

Type genus: *Sinomena* Zeng, Chen et Zhang(gen. nov.)

Diagnosis: Strophomenoid with medium shell; ornamentation variably developed; semicircular in outline; planoconvex to gently concavoconvex in lateral profile; with dental plates which are short and wide, and there bears development of dental plate denticulates, but no hinge line denticulates; widely divergent inner socket ridges with development of denticles on inner socket ridge; cardinal process is bilobed to trilobed; sometimes, cardinal process pit and dorsal median ridge are presented.

Genera assigned: *Aphanomena* Bergström(1968), *Eostropheodonta* Bancroft(1949), *Yichangomena* (gen. nov.), *Sinomena*(gen. nov.)and *Hubeinomena*(gen. nov.).

Geological age: From Middle Hirnantian(Late Ordovician) to Llandoverian(Silurian).

Genus: *Yichangomena* Zeng, Zhang et Han(gen. nov.)

Genotype: *Yichangomena dingjiapoensis* Zeng, Zhang et Han(get. et sp. nov.)

Diagnosis: The outline is semicircular; planoconvex in lateral profile; hinge line straight, nearly equal to the greatest width of the shell. The surface marked by strongly radial costellae, increasing by intercalation; concentric filaes are very fine. Shell substance is pseudopunctate.

Ventral interior: The dental plates are short and wide with posterior grooves that are backward oblique extension. From the dental plates to two sides of the delthyrium, all bears a row of 15～16 dental plate denticulates and subparallel to side margin of the delthyrium(see text pl. 52, figs. 1 - 2; illustr. 15 - A). The ventral muscle field is not clear or faintly impressed.

Dorsal interior: The sockets are small and triangular; the inner socket ridges are strong and widely displayed; socket - facing surface of each inner socket ridge bears a row of 3～5 small denticles; trilobed cardinal process, but median cardinal process is fine; the cardinal process pit is deep and wide in front of the trilobed cardinal process; dorsal muscle area is not visible; no dorsal median ridge.

Discussion: The external shell forms, the ornamentation, and the presence dental plate denticulates and denticles——all indicate that the affinities of the new genus *Yichangomena* may lie with new family Sinomenidae. However, it is distinct from any of the other described genera of the new family Sinomenidae in being a row dental plate denticulates from dental plates to each side of the delthyrium, and having a large cardinal process pit in front of the trilobed process.

Range and distribution: Middle Hirnantian(Late Ordovician); Central China.

Genus *Sinomena* Zeng, Chen et Zhang(gen. nov.)

Genotype: *Sinomena typica* Zeng, Chen et Zhang(gen. et sp. nov.)

Diagnosis: The shells are medium, semicircular in outline, almost planoconvex to gently concavoconvex in lateral profile; hinge - line straight, slightly narrower than the greatest width of the shell. The surface is marked by radial costellae, increasing by intercalation and occasionally by bifurcation; weakly concentric filae. Shell substance is pseudopunctate.

Ventral interior: The dental plates are short and wide. On the inner side of the dental plate, there bears a row of 4 - 6 dental plate denticulates, subvertical to inside margin of the dental plate(see text pl. 53,fig. 1;illusrt. 16 - A). The ventral muscle field is obscure.

Dorsal interior: The socket ridges are wide, and widely divergent. The socket facing surface of each socket ridge bears a row of 4 - 5 denticles (see text pl. 12, fig. 3c; illustr. 16 - B). The cardinal process is trilobed. A very short, high dorsal median ridge extends forward from a point front of the median process, and sometimes, that ankylosed with the median process(see text pl. 12, fig. 4b). The dorsal muscle area does not exist.

Comparison: The new genus *Sinomena* is similar to *Aphanomena* Bergström(1968), but it can be easily distinguished from the latter by its very short, high dorsal median ridge and a row dental plate denticulates subvertical to inside margin of the dental plate. Whereas, the latter is not dorsal median ridge, and dental plate denticulates are subparallel to the inside margin of the dental plate. This new genus differs from *Yichangomena* (gen. nov.). The former, in front of the trilobed cardinal process, bears a short, high dorsal median ridge, but the *Yichangomena* has a large cardinal process pit, no dorsal median ridge.

Range and distribution: Late Ordovician(Middle Hirnantian); Central China.

Genus *Hubeinomena* Zeng, Chen et Zhang(gen. nov.)

Genotype: *Hubeinomena* wangjiawanensis Zeng, Chen et Zhang(gen. et sp. nov.)

Diagnosis: The shells are medium, semicircular in outline; from planoconvex to slightly concavo-convex in lateral profile; hinge - line straight, subequal to the maximum valve width. Surface is marked by round costae, increasing by bifurcation and occasionally by intercalation; concentric filaes are very fine. Shell substance is pseudopunctate.

Ventral interior: The hinge teeth are slender ridge - shaped; dental plates are short and wide, on the surface of each dental plate, there bears 2 short, wide denticulates, slightly oblique to inside margin of the dental plate. The ventral muscle field is obscure or faintly impressed.

Dorsal interior: The cardinalia is very wide, short; socket narrow hole - shaped; inner socket ridge extremely splay; on the surface of each inner socket ridge, there bears a row of 2 - 3 denticles; densely bilobed cardinal process, short and thick. The dorsal muscle field is obscur, sometimes is visible, and is invert bottle in shape(see text pl. 54, fig. 2).

Comparison: The external shell forms and interior structures of new genus *Hubeinomena* are somewhat similar to *Eostropheodonta* Bancroft(1949), but it can be easily distinguished from the latter by its round costae, increasing by bifurcation and occasionally by intercalation; concentric filaes are very fine; bears 2 short, wide denticulates on each dental plate, slightly oblique to inside margin of the dental plate; bilobed cardinal process is very short, thick. However, in *Eostropheodonta* the costellae are very fine or fascicostellae, dental plate denticulates are in posterior part of the dental plate, subvertical to the posterior groove(see text pl. 50, fig. 1 illustr. 14 - A).

Range and distribution: Late Ordovician(Middle Hirnantian); Central China.

Superfamily Plectambonitoidea Jones, 1928
   Family Xenambonitidae Cooper, 1956
     Subfamily Aegiromeninae Havlíček, 1961

Genus *Trimena* Zeng, Wang et Peng(gen. nov.)

Genotype: *Trimena wangjiawanensis* Zeng, Wang et peng(gen. et sp. nov.)

Diagnosis: The shells are small, semicircular in outline, plano – to slightly concavoconvex in lateral profile; hinge line straight, equal to the greatest width of the shell. Surface marked by paucicostellae, round costellae, increasing by intercalation. Shell substance is pseudopunctate.

Ventral interior: The dental plates are short, widely divergent. The ventral muscle field is prominent; diductor scars are elongatly elliptical, widely splayed. The median myophragm is short in height. The adductor is not clear.

Dorsal interior: The inner socket ridges are short and wide, strongly divergent. The cardinal process is bilobed that fused directly to inner socket ridges. The cardinal process pit is small in front of the bilobed cardinal process. The dorsal median septum and lateral septa are quite prominent, but the bema is faint.

Comparison: This new genus is similar to *Aegiria* Öpik(1933) and *Aegiromena* Havliček(1961). It differs from *Aegiria* in being dorsal median septum and lateral septa, its bilobed ventral muscle fields are widely divergent; however, *Aegiria* only has a dorsal median septum. It differs from *Aegiromena* in bearing dorsal median septum and lateral septa and widely divergent bilobed ventral muscle field; whereas, the latter only has two dorsal septa, and its ventral muscle fields are short, narrowly bilobed.

Range and distribution: Late Ordovician (Middle Hirnantian); Central China.

Superfamily Dalmanelloidea Schuchert, 1913
Family Dalmanellidae Schuchert, 1913
Subfamily Dalmanellinae Schuchert, 1913

Genus *Paramirorthis* Zeng, Wang et Peng (gen. nov.)

Genotype: *Paramirorthis minuta* Zeng, Wang et Peng (gen. et sp. nov.)

Diagnosis: The shells are very small, transversely semicircular in outline, gently ventribi-convex in profile; straight hinge line, slightly narrower than the greatest width of the shell. The anterior commissure is rectimarginate. Surface is marked by paucicostellae, simple, edges ridgeshaped costellae, is usually unbifurcate or occasionally bifurcate. Shell substance is fine punctate.

Ventral interior: The dental plates are short and thin, widely displayed. The ventral muscle field is not clear.

Dorsal interior: The cardinalia is very wide and short. The sockets are prominent, slender groove-shaped; brachiophore is plate – like, widely displayed. The cardinal process is very small, knob – shaped in posterior part of the notothyrial cavity, but sometimes is a simple edge. The brachiophore supports are very short, slightly incurved or slightly converged, sometimes approximately parallel. The notothyrial cavity is wide and deep, subcircular in outline. The dorsal muscle field is not clear with no dorsal median septum.

Comparison: The new genus *Paramirorthis* is most closely related to genus *Mirorthis* Zeng (1983), but it can be distinguished from the *Mirorthis* by following differences: The shells of this new genus are very small, transversely semicircular in outline. The surface, marked by paucicostellae, is usually unbifurcate or occasionally bifurcate. The dental plates are short and thin, widely displayed. The cardinalia is very wide and very short; brachiophore is plate – like, widely displayed; having a knob – shaped cardinal process; brachiophore supports are very short, slightly incurved or convergent; notothyrial cavity is wide, subcircular and deep. Whereas, *Mirorthis* has a subcircular in outline; ornamentation is multicostellae. The cardinalia is narrow, bears inner and outer brachiophore supports; notothyrial cavity is narrow and long.

Range and distribution: Late Ordovician(Middle Hirnantian); Central China, Poland, ? Britain.

Superfamily Dalmanelloidea Schuchert, 1913
    Family Dysprosorthidae Zeng et Zhang(fam. nov.)

Type genus: *Dysprosorthis* Rong, 1984

Diagnosis: Dalmanelloid with small shell, subequal ventribiconvex, gently ventral median fold and wide dorsal median sulcus; typically fascicostellae ornamentation. Shell substance is fine endopunctate.

Ventral interior: The dental plates are short and fine; ventral muscle field is suboval.

Dorsal interior: The cardinalia is very atypical, very short and very wide; hinge sockets are slender, and lie in the posterior sides of the hinge line, widely divergent in anteriorly lateral areas of the notothrium; brachiophores are thick ridge – shaped, subparallel to side margin of the notothyrium; usually no brachiophore supports, but very rudimental in old shells; the notothyrium is very short and wide, triangular; cardinal process is single, short.

Discussion: The genus *Dysprosorthis* Rong (1984) was assigned to Chrustenoridae by Harper (2000), but it must be pointed out that to be a most unusual Dalmanelloid, this genus is characterized by distinct cardinalia, small shell, typically fassicostellate ornamentation, gently ventral fold and wide dorsal sulcus. These characteristics are so distinctive that the Dysprosorthis Rong cannot be placed with confidence in any existing Dalmanelloid family. The new family Dysprosorthidae, therefore, is herein established to accommodate the single genus.

Geological range: Middle Hirnantian(Late Ordovician); Central China.

5.2 Some new materials of brachiopods in Luorepinian

Suborder Orthidina Schuchert et Cooper, 1932
    Superfamily Plectorthoidea Schuchert et LeVene, 1929
        Family Toxorthidae Rong, 1984

            Genus *Minutorthis* Zeng, Chen et Zhang(gen. nov.)

Genotype: *Salopina? yichangensis* Rong et Yang, 1981

Diagnosis: Orthoid with very small shell, semicircular in outline, planoconvex in lateral profile; straight hinge line, equal to the greatest width of the shell. The cardinal extremities are acutely angular. The ventral valve is moderately convex, and with the greatest convexity in the unbonal region. The ventral interarea is short, apsacline. The dorsal valve is very thin and even, and there is a low knob – like in front of the dorsal beak; dorsal interarea is shorter than the ventral one, anacline. The ornamentation is paucicostellae. Shell substance is impunctate.

Ventral interior: The hinge teeth are small, elliptical ridge – shaped; dental plates are short and small, widely displayed. The ventral muscle field is slightly bilobed, diductor scars are slightly longer than the adductor one; no ventral median septum.

Dorsal interio: The cardinalia is very wide; hinge sockets are small, elliptical pit – shaped; brachiophores are wide plate – shaped, widely displayed; loss of brachiophore supports; single cardinal process; dorsal median septum is very low and faint; dorsal muscle is obscure.

Comparison: This new genus is similar to *Toxorthis* Temple(1968), but it can be easily distinguished from the latter by its semicircular in outline, planoconvex in lateral profile, paucicostellae ornamentation, single cardinal process; whereas, *Toxorthis* has greatly transversely elongated outline, ventribiconvex shell, paucicostae, bilobed cardinal process.

Range and distribution: Late Aeronian(Silurian); Central China

Superfamily Pentameroidea M′Coy, 1844
Family Pentameridae M′Coy, 1844

Genus *Centreplicatus* Zeng, Zhang, et Han(gen. nov.)

Genotype: *Centreplicatus triangulatus* Zeng, Zhang et Han(gen. et sp. nov.)

Diagnosis: The shells are medium size, subtriangular in outline; non–trilobate; moderately ventribiconvex in lateral profile; anterior commissure subrectimarginate or slightly sulcate, with a strongly incurved ventral beak. The surfaces bear faint, low, wide symmetrically radiating plications only restricted to longitudinal part of the both valves; whereas, the both lateral areas are smooth.

In the ventral valve, there is a very long spondyliun, about four–fifths of the length of ventral valve; median septum about two–fifths of the length ventral of valve.

In the dorsal valve, the brachial apparatus composed of long, subparallel inner and outer plates, about one half length of dorsal valve; but, anterior parts of the inner plates are incurved and join smoothly with one another in U shape.

Comparison: The external shell forms and interior structures indicate that the affinities of the new genus *Centreplicatus* may lie with Pentameridae M′Coy(1844). It differs from other members of Pentameridae in that the velves of this new genus are medium size, subtrigonal in outline; the surfaces there bear three low and wide, symmestrically radial plications that only restricted to lingitudinal part of each valve; in the dorsal valve there have two long, subparallel inner plates that are incurved in anterior parts, and join smoothly with one another in U shape.

Range and distribution: Late Aeronian(Silurian); Centrel China.

Genus *Apopentamerus* Boucot et Johnson, 1979

Subgenus *Apopentamerus* (*Enclosurus*) Zeng, Wang et Peng(subgen. nov.)

Subgenotype: *Apopentamerus fenxiangensis* Zeng, 1987

Diagnosis: The shells are large, smooth, non–trilobate, subpyriform in outline, moderately ventribiconvex in lateral profile. The anterior commissure is rectimarginate. The spondylium, ventral median septum, inner and outer plates etc. structurs, all look *Apopentamerus*, but those new subgenus having unusually calcific enclosures in the visceral cavity.

Comparison: This new subgenus *Apopentamerus*(*Enclosurus*) differs from *Apopentamerus*(*Apopentamerus*) in having a unusually calcific enclosures in the visceral cavity; whereas the latter is not calcific enclosures.

Range and distribution: Late Aeronian(Silurian); Central China.

索 引

属和种（Genera and species）	图版（pl.）	图（fig.）	页（P.）
Acanthocrania *A. yichangensis*	1	8a – 8d	20
Aegiria *A. planissima*	16 17	1 – 12 1 – 12	41
Aegiromena *A. diplosepta*(sp. nov.)	19	3 – 9	39
Aphanomena *A. parvicostellata*	9 10 51	1 – 8 1 – 10 1	32
Apopentamerus(*Enclosurus*)(subgen. nov.) *A.*(*E.*)*fenxiangensis*	68	1 – 3	96
Centreplicatus(gen. nov.) *C. triangulatus*(gen. et sp. nov.)	66 67	1 – 4 1 – 5	94
Chonetoidea *C. simplex*(sp. nov.)	18 19	1 – 12 1, 2	44
Cliftonia *C. elongata*(sp. nov.)	45	1 – 5	52
C. oxoplecioides	22 56	1 – 12 1, 2	51
Coolinia *Coolinia* sp.	20	6, 7	48
Craniops *C. partibilis*	1	4 – 7	19
Dalmanella *D. testudinaria*	23 24	1 – 11 1 – 12	55
Draborthis *D. caelebs*	42 43	1 – 12 1 – 11	78
Drabovia *D. dingjiapoensis*(sp. nov.)	32 33	1 – 10 1 – 8	65

属和种（Genera and species）	图版(pl.)	图(fig.)	页(P.)
D. ? sp.	34	1	66
Drabovinella 　　*D. yichangensis*(sp. nov.)	34	2-5	67
Dysprosorthis 　　*D. sinensis*	44 55	1-9 1,2	81
Eostropheodonta 　　*E. hirnantensis*	7 8 50	1-12 1-11 1,2	33
Fardenia 　　*F. scotica*	20	5,8,9,11	47
Hindella 　　*H. crassa incipiens*	45 46 58	6-11 1-10 1-3	85
H. ? *elegans*(sp. nov.)	46 47	11 1-6	87
Hirnantia 　　*H. fecunda*	34	6-12	74
H. magna	36	3-12	71
H. sagittifera morph. Bohemia	35 36	1-12 1,2	69
H. sagittifera morph. Poland	37	1-11	70
H. septumis	38	1-12	72
Hubeinomena(gen. nov.) 　　*H. wangjiawanensis*(gen. et sp. nov.)	13 14 54	1-8 1-4 1-3	38
Kinnella 　　*K. kielanae*	39 40 51	1-10 1-8 2	75
K. robusta	41	1-11	77
Leptaena 　　*L. huanghuaensis*	2 3 49	6-10 1-8 1	24

索 引

属和种（Genera and species）	图版（pl.）	图（fig.）	页（P.）
Leptaenopoma *L. rugosas*	5	1-9	27
L. trifidum	4 49	1-9 2	26
L. yichangense	6	1-9	28
Leptellina（*Merciella*） *L.*（*M.*）*striata*	64	1-3	88
Leptelloidea *L. silurica*（sp. nov.）	64	4-7	89
Minutomena（gen. nov.） *M. yichangensis*（gen. et sp. nov.）	2	1-5	22
Mirorthis *M. mira*	31 57	5-11 2	62
Minutorthis（gen. nov.） *M. yichangensis*	65	6a-6b	91
Onniella *O. yichangensis*	25 26	1-12 1-11	57
Orbiculoidea *O.? sp.*	1	1-3	18
Paramirorthis（gen. nov.） *P. minuta*（gen. et sp. nov.）	29 30 31 57	6-11 1-11 1-4 1	63
Paromalomena *P. polonica*	14 15	5-11 1-12	29
Philhedra *P. sp.*	1	9-11	21
Plectothyrella *P. crassicosta*	47 48 60	7-9 1-10 1,2	82
Sinomena（gen. nov.） *S. typica*（gen. et sp. nov.）	12 53	1-9 1,2	37
Sulcipentamerus *S. dorsoplanus*	65	1-5	92

属和种(Genera and species)	图版(pl.)	图(fig.)	页(P.)
Toxorthis			
T. mirabilis	33	9-11	54
Trimena(gen. nov.)	19	10	46
T. wangjiawanensis(gen. et sp. nov.)	20	1-4	
Triplesia			
T. fenxiangensis	21	7-12	50
T. yichangensis	20	10	49
	21	1-6	
Trucizetina	27	1-12	59
T. yichangensis	28	1-11	
T. ? parallela(sp. nov.)	29	1-5	60
Yichangomena(gen. nov.)	11	1-11	35
Y. dingjiapoensis(gen. et sp. nov.)	52	1,2	

图版及图版说明

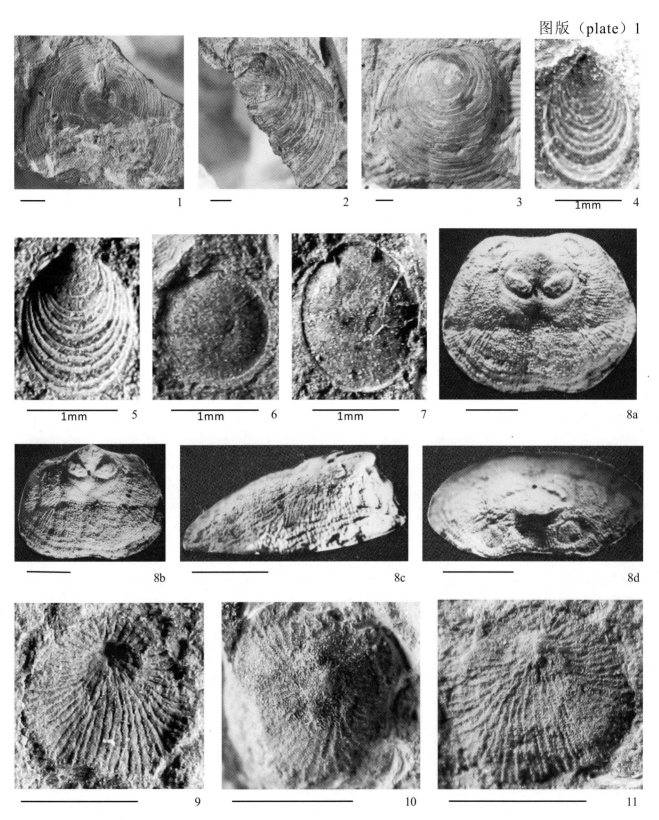

图版1,除加注外,所有比例线条=2mm——plate 1(all scale bars=2mm for all exceptions as noted)

1-3. *Obiculoidea*? sp.

    1. Ventral internal mold, WH3, HB710; 2. Ventral internal mold, WH2, HB699; 3. Dorsal exterior?, DH2, HB351.

4-7. *Craniops partibilis* (Rong)

    4. Ventral external mold, WH1, HB692; 5. Ventral external mold, WH1, HB72; 6. Ventral internal mold, WH1, HB71; 7. dorsal internal mold, WH2, HB216.

8a-8d. *Acanthocrania yichangensis* Zeng Respectively dorsal, anterior, side and posterior views of dorsal internal mold of same specimen, HK1, IV45740.

9-11. *Philhedra* sp.

    9. Dorsal external mold, WH3, HB684; 10. Dorsal exterior, WH2, HB700; 11. Dorsal exterior, DH2, HB352.

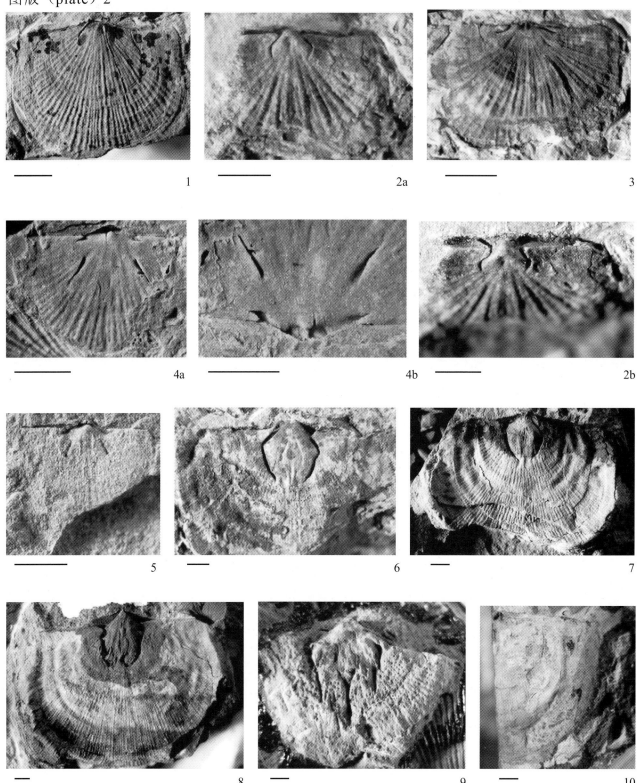

图版 2 所有比例线条＝2mm——plate 2(all scale bars＝2mm)

1-5. *Minutomena yichangensis* gen. et sp. nov.

1. Ventral internal mold, paratype, WH2, HB677; 2a-2b. Ventral and posterior views of ventral internal mold of same specimen, paratype, Pm065-6-1F, YB1; 3. Dorsal internal mold, paratype, pm065-6-1F, YB2; 4a-4b. Dorsal and posterior views of dorsal internal mold of same specimen, holotype, DH2, HB678; 5. Dorsal internal mold, WH2, HB377.

6-10. *Leptaena huanghuaensis* Zeng

6. Ventral internal mold, WH2, HB69; 7. Ventral internal mold, DH2, HB322; 8. Ventral internal mold, WH2, HB283; 9. Dorsal internal mold, WH2, HB97a; 10. Dorsal internal mold, WH2, HB96.

图版3 所有比例线条＝2mm——plate 3(all scale bars＝2mm)

1–8. *Leptaena huanghuaensis* Zeng

1a–1c. Ventral internal mold of same specimen, DH2, HB357; 1a. ventral view, 1b. ventral interarea and pseudodeltidium, 1c. hinge teeth and dental plates; 2. Ventral internal mold, WH3, HB285; 3. Ventral internal mold, DH2, HB319; 4. Ventral internal mold, WH3, HB706; 5. Bifid cardinal process, DH2, HB304; 6. Dorsal internal mold, HK2, IV45843; 7. Bifid cardinal process, DH3, HB113; 8a–8c. Same specimen, DH3, HB305b; 8a. dorsal external mold, 8b. dorsal internal mold, 8c. bifid cardinal process and secondary sockets.

图版（plate）4

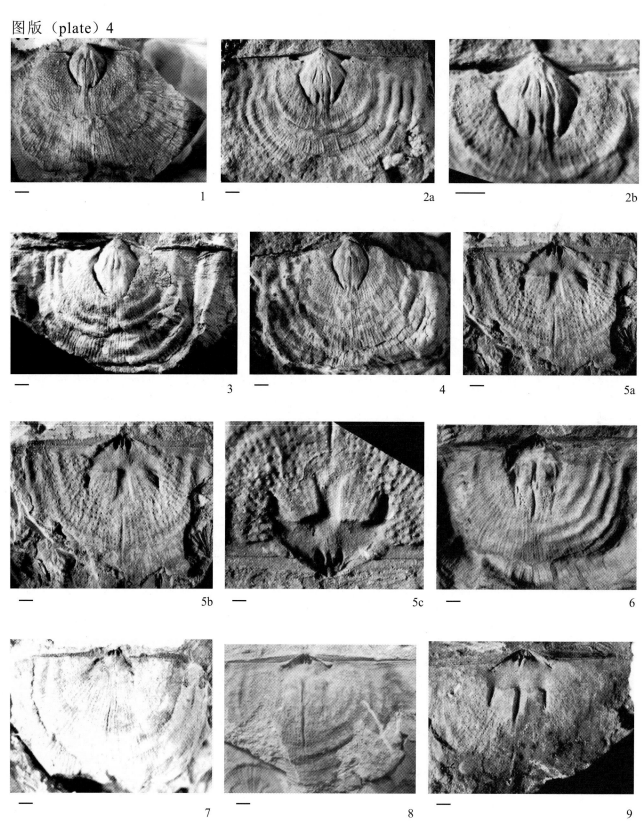

图版 4　所有比例线条＝2mm——plate 4(all scale bars＝2mm)

1－9. *Leptaenpoma trifidum* Marek et Havlíček

　　1. Ventral internal mold, DH2, HB355; 2a－2b. Ventral internal mold, DH2, HB309, muscle area of same specimen; 3. Ventral internal mold, WH2, HB279; 4. Ventral internal mold, DH2; HB292; 5a－5c. Same specimen: dorsal internal mold and its dorsal and posterior views showing strongly dorsal platform and trifided cardinal process, metatype, DH3, HB310; 6. Dorsal internal mold (old specimen), DH3, HB491; 7. Dorsal internal mold, WH3, HB293; 8. Dorsal internal mold, WH2, HB58; 9. Dorsal internal mold, DH3, HB312.

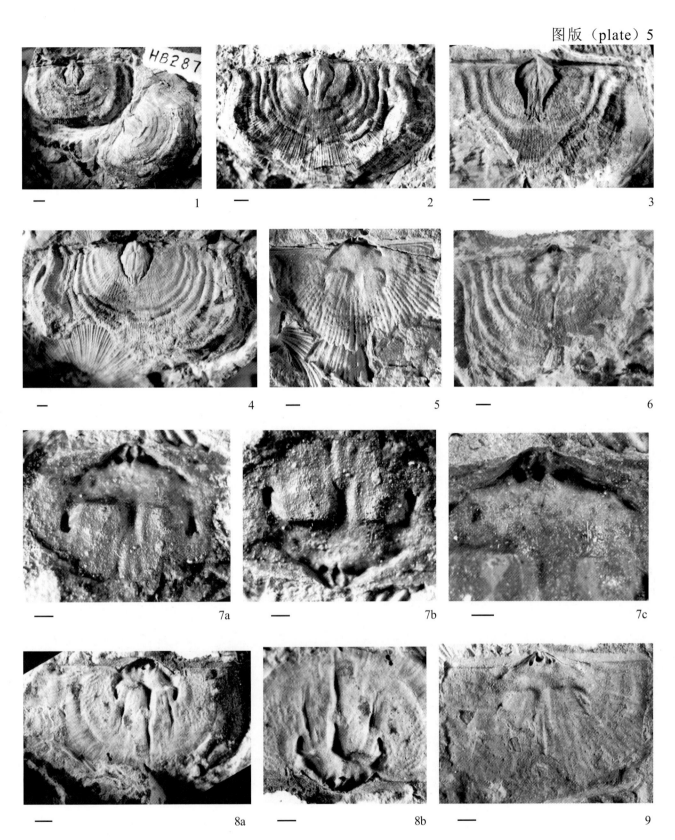

图版 5 所有比例线条=2mm——plate 5(all scale bars=2mm)

1-9. *Leptaenopoma rugosas* Zeng

1. Ventral internal mold, WH2, HB287; 2. Ventral internal mold, WH2, HB277; 3. Ventral internal mold, WH3, HB289; 4. Ventral internal mold, DH2, HB308; 5. Dorsal internal mold, DH3, HB154; 6. Dorsal internal mold, WH2, HB94; 7a-7c. Dorsal internal mold of same specimen, respectively dorsal, posterior, and cardinal process views showing strongly dorsal platform and trifided cardinal process, metatype, DH2, HB307; 8a-8b. Dorsal internal mold(old specimen) of same specimen, dorsal and posterior views, WH2, HB323; 9. Dorsal internal mold, DH3, HB314.

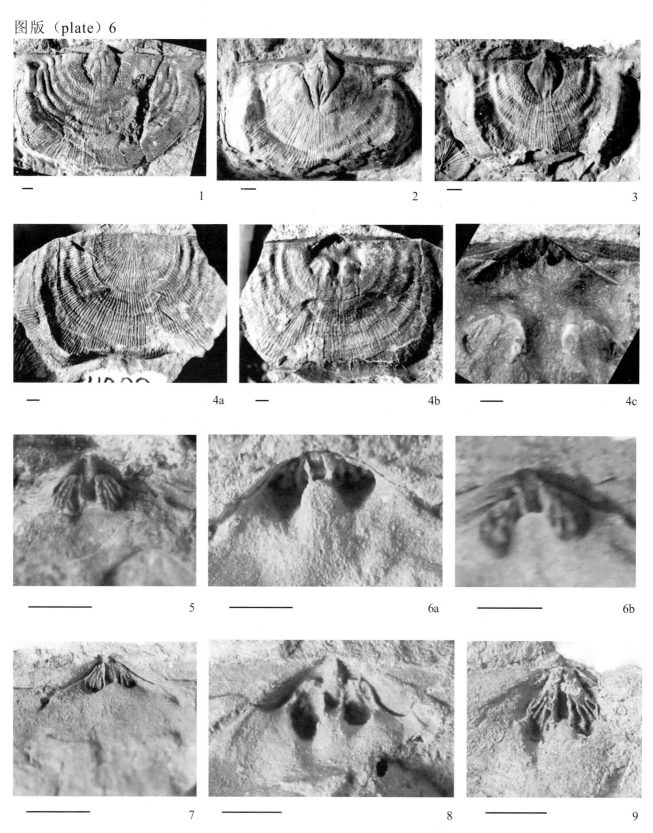

图版 6 所有比例线条＝2mm——plate 6(all scale bars＝2mm)

1-9. *Leptaenopoma yichangense* Zeng

1. Ventral internal mold, WH3, HB291; 2. Ventral internal mold, DH2, HB311; 3. Ventral internal mold, WH2, HB386; 4a-4c. Same specimen: 4a. dorsal external mold, 4b. dorsal internal mold, 4c. details of cardinalia, DH2, HB92; 5. Details of cardinalia, WH2, HB59; 6a-6b. Same specimen showing details of cardinalia, DH3, HB110; 7. Details of cardinalia, WH2, HB610; 8. Cardinal process, WH2, HB281; 9. Cardinal process, WH2, HB280.

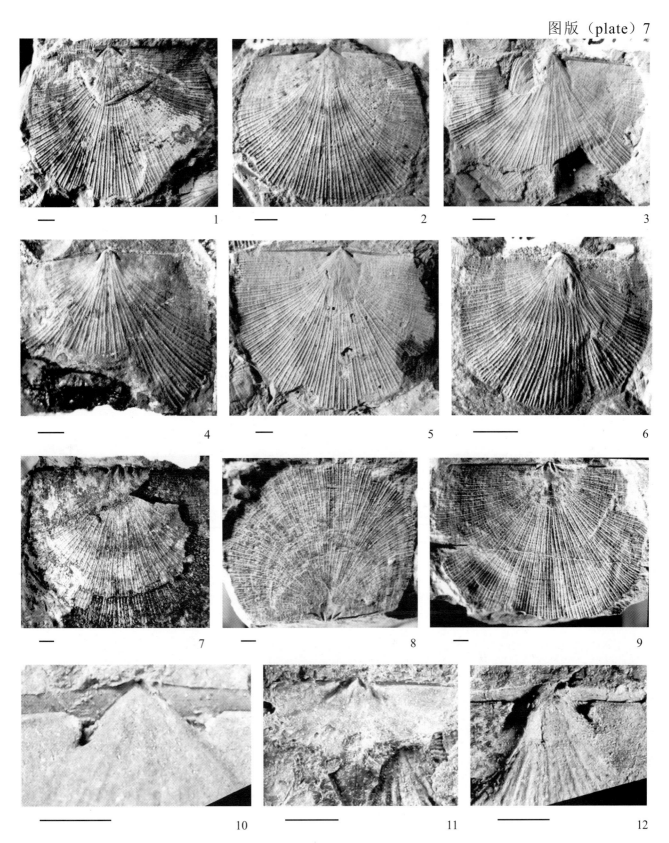

图版7 所有比例线条＝2mm——plate 7(all scale bars＝2mm)

1–12. *Eostropheodonta hirnantensis* (M'Coy)

1. Ventral internal mold, WH3, HB584; 2. Ventral internal mold, DH2, HB346; 3. Ventral internal mold, WH2, HB707; 4. Ventral internal mold, WH3, HB569; 5. Ventral internal mold, WH2, HB384; 6. Ventral internal mold, WH2, HB28; 7. Dorsal internal mold, WH3, HB40; 8. Posterior view of dorsal internal mold, WH2, HB593; 9. Dorsal internal mold, WH2, HB592; 10. Ventral internal mold, showing dental plate denticulates, DH2, HB689; 11. Dorsal internal mold, DH2, HB598; 12. Dental plate denticulates, WH2, HB385.

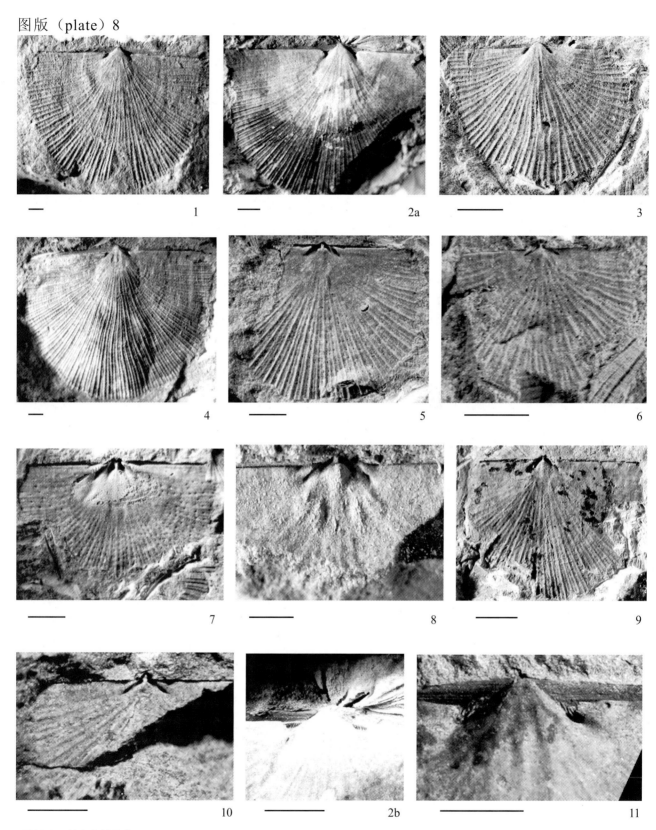

图版 8 所有比例线条=2mm——plate 8(all scale bars=2mm)

1–11. *Eostropheodonta hirnantensis* (M'Coy)

1. Ventral internal mold, WH2, HB27; 2a–2b. Same specimen: Ventral internal mold and details of dental plate denticulates, DH2, HB536; 3. Ventral internal mold, WH3, HB585; 4. Ventral internal mold, WH1, HB388; 5. Dorsal internal mold, DH3, HB496; 6. Dorsal internal mold, WH2, HB451; 7. Dorsal internal mold, DH2, HB89; 8. Dorsal internal mold, WH3, HB600; 9. Dorsal internal mold, WH2, HB578; 10. Dorsal internal mold, WH2, HB390; 11. Details of dental plate denticulates, WH3, HB40.

图版（plate）9

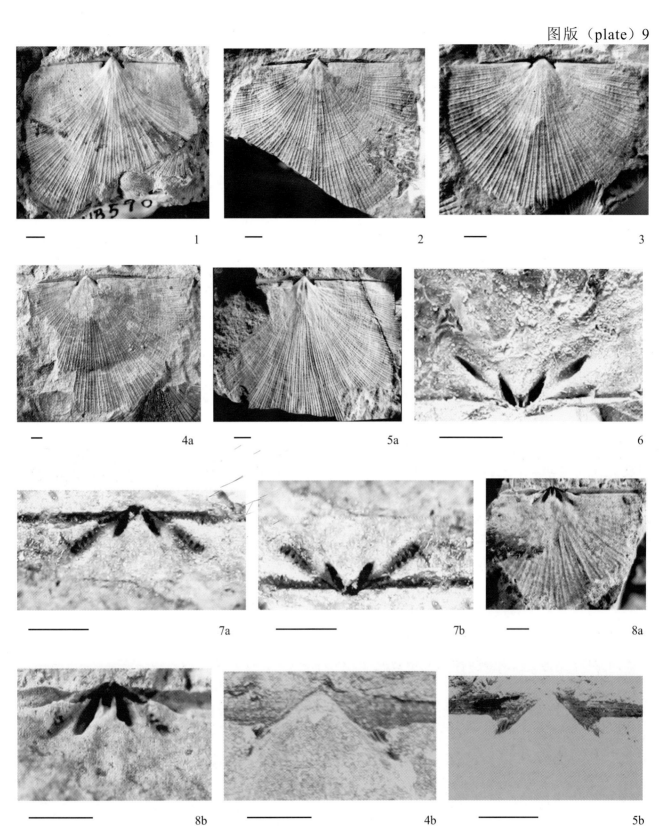

图版 9 所有比例线条=2mm——plate 9(all scale bars=2mm)

1-8. *Aphanomena parvicostellata* Rong

1. Ventral internal mold, WH3, HB590; 2. Ventral internal mold, WH1, HB485; 3. Ventral internal mold, DH2, HB360; 4a-4b. Same specimen: ventral internal mold and dental plate denticulates, WH2, HB577; 5a-5b. Same specimen: ventral internal mold and dental plate denticulates, WH2, HB589; 6. Posterior view of cardinalia, WH2, HB59; 7a-7b. Same specimen: dorsal and posterior views of cardinalia showing development denticles on inner socket ridge, DH2, HB539; 8a-8b. Same specimen showing cardinalia and denticles on inner socket ridge, DH2, HB442.

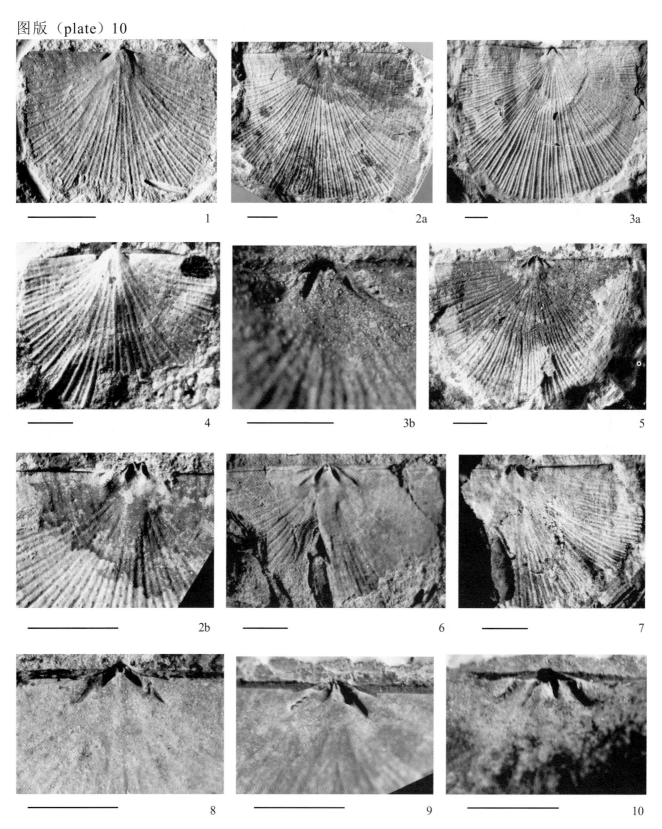

图版 10 所有比例线条＝2mm——plate 10(all scale bars＝2mm)

1-10. *Aphanomena parvicostellata* Rong

1. Ventral internal mold, WH2, HB41; 2a-2b. Same specimen; dorsal internal mold and showing details of cardinalia, WH2, HB708; 3a-3b. Same specimen; dorsal internal mold and enlargements of cardinalia showing denticles on inner socket ridge, WH1, HB488; 4. Ventral internal mold, DH2, HB347; 5. Dorsal internal mold, WH2, HB44; 6. Dorsal internal mold, WH2, HB580; 7. Dorsal internal mold, WH1, HB489; 8. Dorsal internal mold, WH2, HB582; 9. Denticles on inner socket ridge, WH1, HB596; 10. Denticles on inner socket ridge, DH2, HB690.

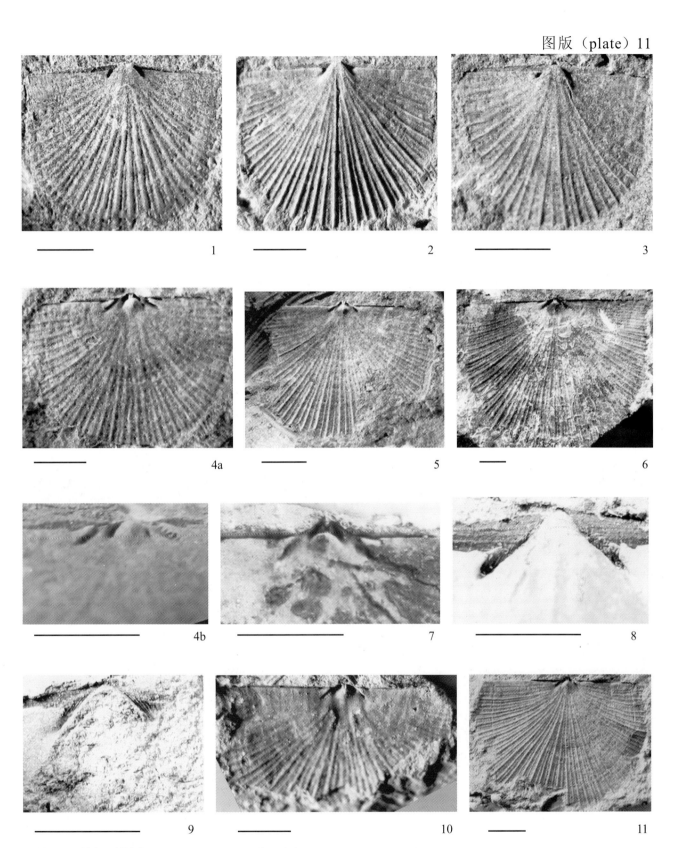

图版 11 所有比例线条＝2mm——plate 11(all scale bars＝2mm)

1-11. *Yichangomena dingjiapoensis* gen. et sp. nov.

1. Ventral internal mold, WH1, HB100; 2. Ventral internal mold, paratype, DH2, HB345; 3. Ventral internal mold, WH3, HB372; 4a-4b. Same specimen: dorsal internal mold, holotype, and enlargements of cardinalia showing cardinal process pit and denticles of inner socket ridge, DH2, HB519; 5. Dorsal internal mold, DH2, HB510; 6. Dorsal internal mold, paratype, WH3, HB568; 7. Details of cardinalia, WH3, HB604; 8. Details of dental plate denticulates, paratype, DH2, HB544; 9. Details of dental plate denticulates, paratype, DH2, HB520; 10. Dorsal internal mold, WH1, HB490; 11. Dorsal internal mold, paratype, WH2, HB545.

图版（plate）12

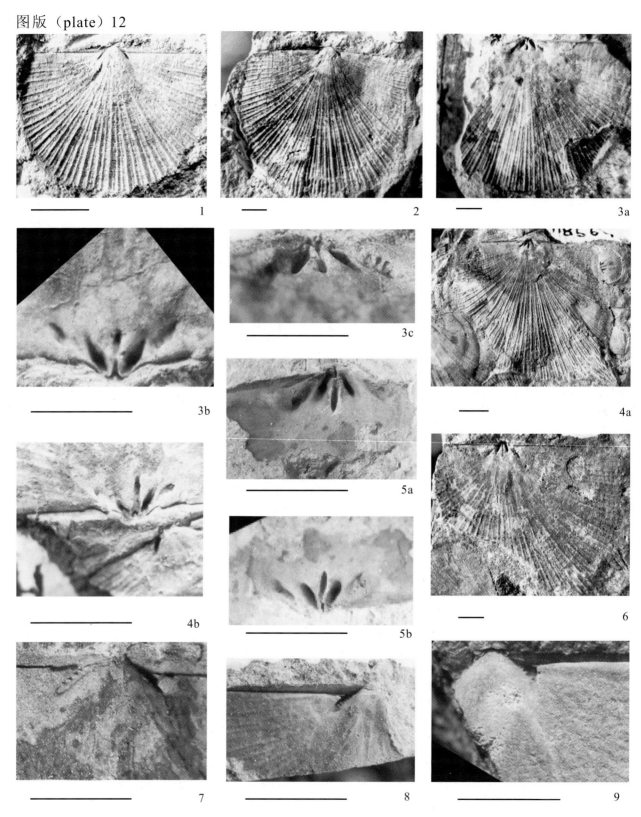

图版 12　所有比例线条＝2mm——plate 12(all scale bars＝2mm)

1－9. *Sinomena typica* gen. et sp. nov.

　　1. Ventral internal mold, WH3, HB588; 2. Ventral internal mold, paratype, WH2, HB606; 3a－3c. Same specimen, holotype; 3a. dorsal internal mold, 3b. posterior view of cardinalia, 3c. Denticles on inner socket ridge, WH2, HB563; 4a－4b. Same specimen; dorsal internal mold and its posterior view showing dorsal median ridge, WH2, HB564; 5a－5b. Same specimen, dorsal and posterior views of dorsal internal mold showing cardinalia and dorsal median ridge, DH3, HB484; 6. Dorsal internal mold, WH3, HB567; 7. Dental plate denticulates, paratype, WH2, HB288; 8. Dental plate denticulates, paratype, WH2, HB607; 9. Dental plate denticulates, DH2, HB521.

图版（plate）13

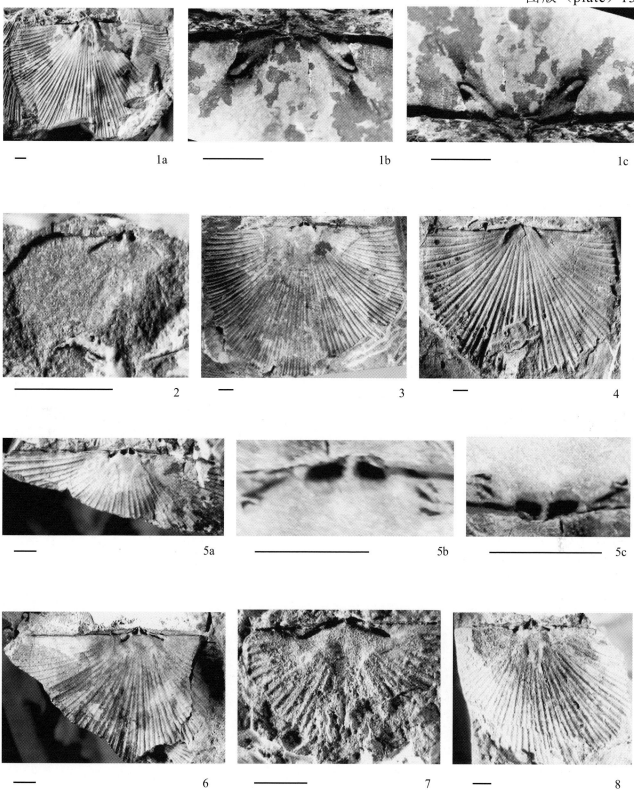

图版 13 所有比例线条＝2mm——plate 13(all scale bars＝2mm)

1-8. *Hubeinomena wangjiawanensis* gen. et sp. nov.

1a-1c. Same specimen: ventral internal mold and its ventral and posterior views of dental plate denticulates, paratype, WH3, HB697; 2. Dorsa internal mold, WH2, HB622; 3. Dorsal internal mold, holotype, WH3, HB556; 4. Ventral internal mold, WH3, HB133; 5a-5c. Dorsal internal mold and its dorsal and posterior views of denticles on inner socket ridge, paratype, WH3, HB696; 6. Dorsal internal mold, WH3, HB660; 7. Dorsal internal mold, WH2, HB468; 8. Dorsal internal mold, WH3, HB453.

图版 14 所有比例线条＝2mm——plate 14(all scale bars＝2mm)

1-4. *Hubeinomena wangjiawanensis* gen. et sp. nov.

    1a-1b. Same specimen;dorsal internal mold and external mold of dorsal valve? and its local enlargements showing details of ornamentation,WH3,HB137;2. Ventral internal mold,WH2,HB31;3. Cardinalia,WH1,HB99;4. Ventral internal mold,HK3,IV45703.

5-11. *Paromalomena polonica* (Temple)

    5. Ventral internal mold,WH3,HB459;6. Ventral internal mold,WH3,HB454;7. Ventral internal mold,WH3,HB455;8. Ventral internal mold,WH2,HB444;9. Ventral internal mold,WH2,HB394;10. Dorsal internal mold,WH2,HB166;11. Dorsal internal mold,WH3,HB467.

图版（plate）15

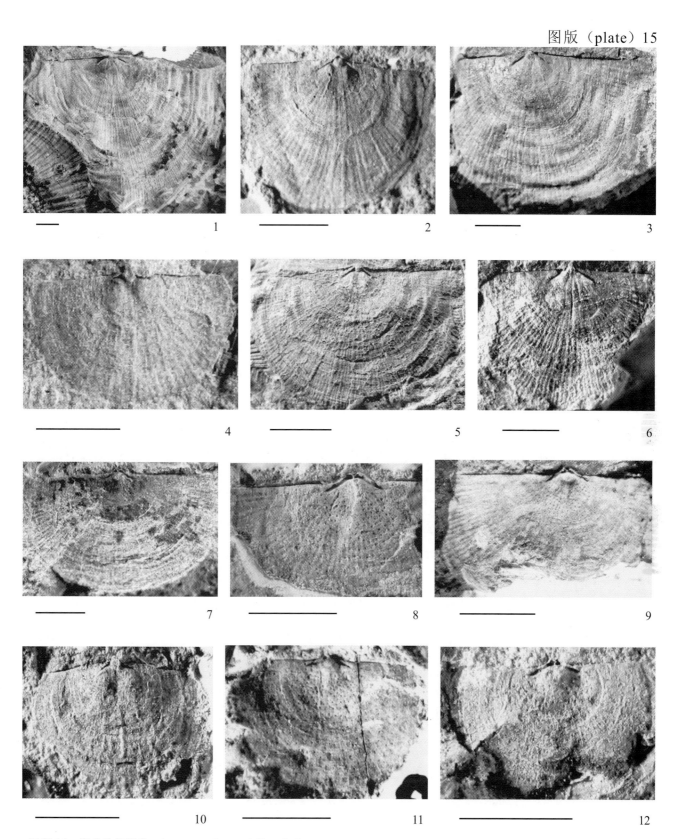

图版 15 所有比例线条=2mm——plate 15(all scale bars=2mm)

1-12. *Paromalomena polonica* (Temple)

1. Ventral internal mold, WH3, HB457; 2. Ventral internal mold, WH2, HB161; 3. Ventral internal mold, WH3, HB456; 4. Ventral internal mold, WH3, HB460; 5. Ventral internal mold, WH3, HB586; 6. Ventral internal mold, DH1, HB87; 7. Dorsal internal mold, WH1, HB446; 8. Dorsal internal mold, WH2, HB561; 9. Dorsal internal mold, DH2, HB713; 10. Ventral internal mold, WH2, HB469; 11. Dorsal internal mold, WH2, HB445; 12. Dorsal internal mold, WH2, HB381.

图版（plate）16

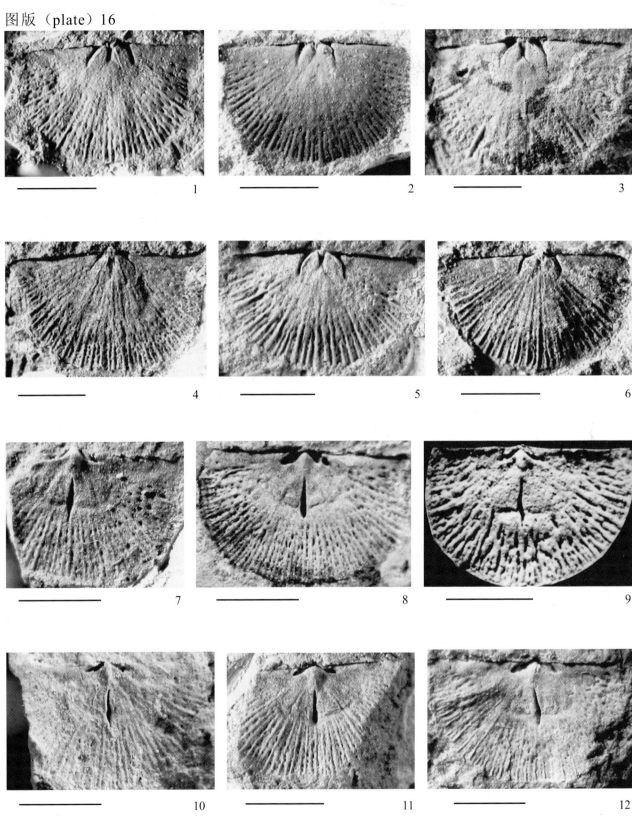

图版 16 所有比例线条＝2mm——plate 16(all scale bars＝2mm)

1-12. *Aegiria planissima* (Reed)

　　1. Ventral internal mold, DH2, HB330; 2. Ventral internal mold, DH3, HB440; 3. Ventral internal mold, DH2, HB427; 4. Ventral internal mold, WH1, HB336; 5. Ventral internal mold, DH2, HB434; 6. Ventral internal mold, WH1, HB389; 7. Dorsal internal mold, DH2, HB423; 8. Dorsal internal mold, DH2, HB415; 9. Dorsal internal mold, HK3, IV45700; 10. Dorsal internal mold, DH2, HB428; 11. Dorsal internal mold, WH2, HB380; 12. Dorsal internal mold, WH2, HB789.

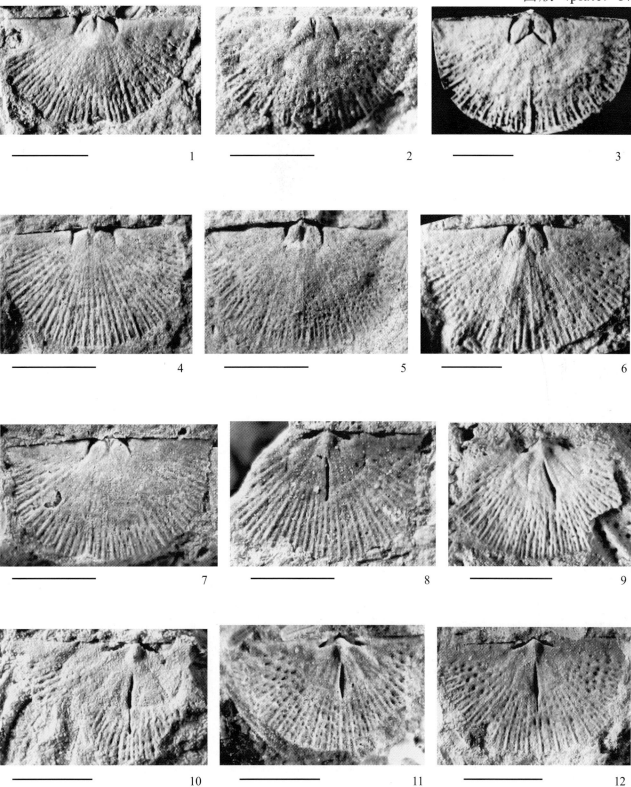

图版 17 所有比例线条＝2mm——plate 17(all scale bars＝2mm)

1-12. *Aegiria planissima* (Reed)

1. Ventral internal mold, DH2, HB359; 2. Ventral internal mold, WH1, HB414; 3. Ventral internal mold, HK1, IV45698; 4. Ventral internal mold, DH2, HB343; 5. Ventral internal mold, DH2, HB439; 6. Ventral internal mold, DH3, HB143; 7. Ventral internal mold, WH2, HB435; 8. Dorsal internal mold, DH2, HB420; 9. Dorsal internal mold, DH2, HB327; 10. Dorsal internal mold, DH3, HB411; 11. Dorsal internal mold, DH2, HB429; 12. Dorsal internal mold, DH2, HB552.

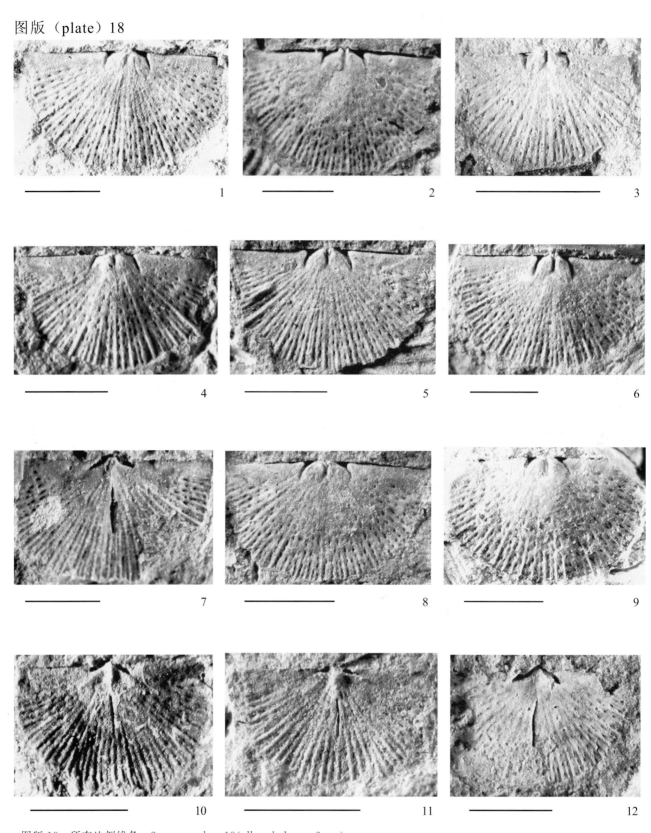

图版 18 所有比例线条＝2mm——plate 18(all scale bars＝2mm)

1-12. *Chonetoidea simplex* sp. nov.

1. Ventral internal mold, DH3, HB493; 2. Ventral internal mold, paratype, DH3, HB574; 3. Ventral internal mold, DH3, HB335; 4. Ventral internal mold, DH2, HB88; 5. Ventral internal mold, paratype, DH2, HB424; 6. Ventral internal mold, paratype, DH2, HB425; 7. Dorsal internal mold, DH2, HB433; 8. Ventral internal mold, DH3, HB449; 9. Ventral internal mold, DH3, HB515; 10. Dorsal internal mold, WH2, HB149; 11. Dorsal internal mold, DH2, HB426; 12. Dorsal internal mold, paratype, DH3, HB470.

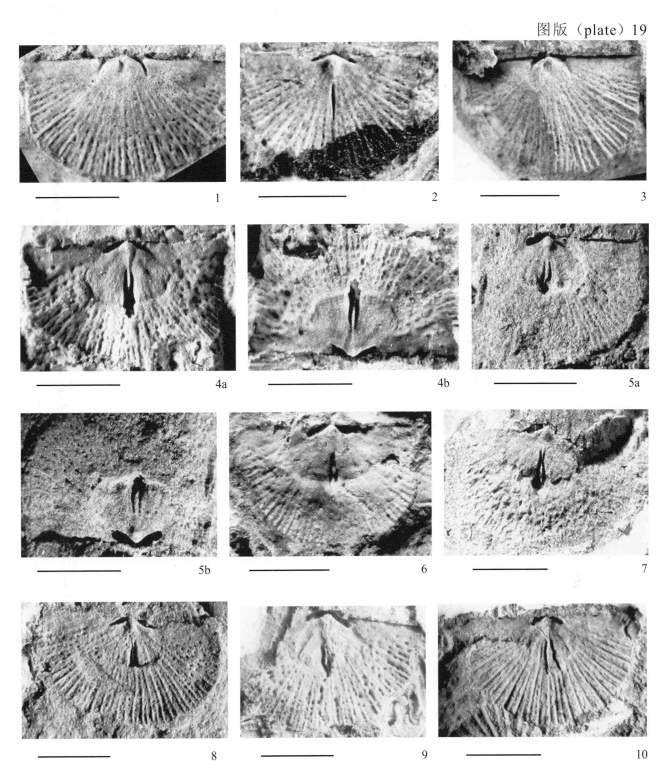

图版 19 所有比例线条=2mm——plate 19(all scale bars=2mm)

1-2. *Chonetoidea simplex* sp. nov.

    1. Ventral internal mold, DH2, HB430; 2. Dorsal internal mold, holotype, DH2, HB431.

3-9. *Aegiromena diplosepta* sp. nov.

    3. Ventral internal mold, paratype, WH2, HB36; 4a-4b. Same specimen: dorsal internal mold and its posterior view showing details of cardinalia, holotype, DH3, HB369; 5a-5b. Same specimen: dorsal internal mold and its posterior view, WH1, HB364; 6. Dorsal internal mold, DH2, HB358; 7. Dorsal internal mold, WH1, HB365; 8. Dorsal internal mold, DH2, HB361; 9. Dorsal internal mold, DH3, HB4367.

10. *Trimena wangjiawanensis* gen. et sp. nov.

    10. Dorsal internal mold, WH3, HB371.

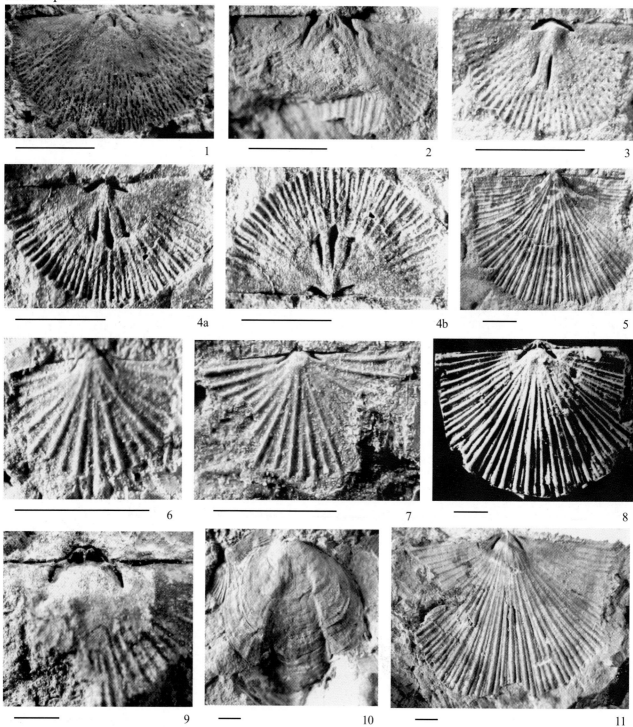

图版 20 所有比例线条＝2mm——plate 20(all scale bars＝2mm)

1-4. *Trimena wangjiawanensis* gen. et sp. nov.
　　1. Ventral internal mold, paratype, DH2, HB537; 2. Ventral internal mold, DH3, HB374; 3. Dorsal internal mold, WH3, HB373; 4a-4b. Same specimen: dorsal internal mold and its posterior view showing details of cardinalia, holotype, WH1, HB370.

5,8-9,11. *Fardenia scotica* Lamont
　　5. Ventral internal mold, WH2, HB29; 8. Dorsal internal mold, HK2, IV45705; 9. Dorsal internal mold, WH2, HB557; 11. Ventral internal mold, WH3, HB294.

6,7. *Coolinia* sp.
　　6. Ventral internal mold, WH2, HB465; 7. Dorsal internal mold, WH3, HB466.

10. *Triplesia yichangensis* Zeng
　　10. Ventral exterior, WH1, HB78.

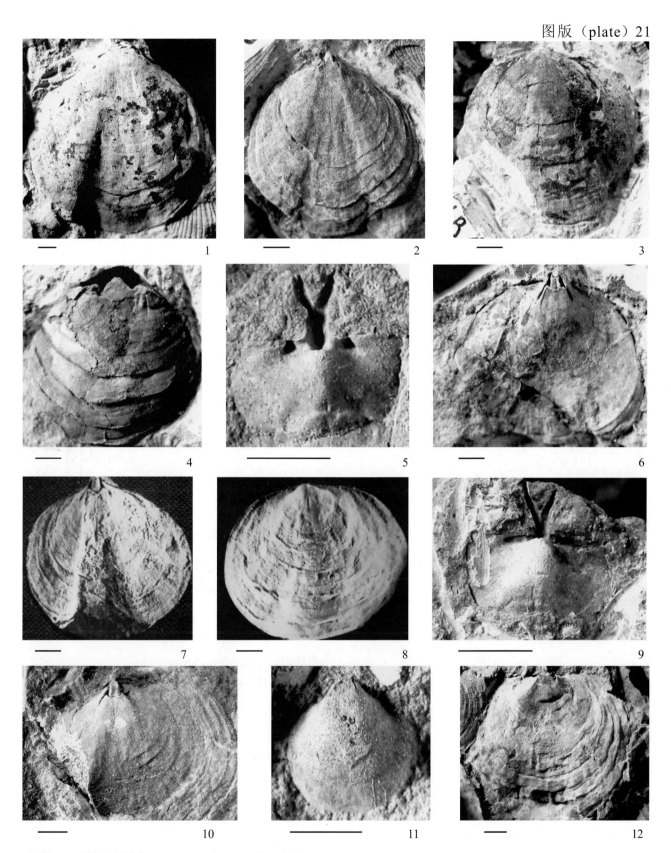

图版 21 所有比例线条＝2mm——plate 21(all scale bars＝2mm)

1-6. *Triplesia yichangensis* Zeng

1. Ventral exterior, WH2, HB46; 2. Ventral internal mold, DH2, HB632; 3. Dorsal exterior, DH2, HB633; 4. Dorsal exterior, DH2, HB93; 5. Dorsal internal mold, DH2, HB634; 6. Ventral internal mold, WH2, HB608.

7-12. *Triplesia fenxiangensis* Yan

7. Ventral internal mold, HH2, HB729; 8. Dorsal exterior, FH2, HB730; 9. Dorsal internal mold, DH2, HB638; 10. Ventral internal mold, DH2, HB637; 11. Ventral internal mold, WH2, HB38; 12. Ventral exterior, WH3, HB651.

图版 22 所有比例线条＝2mm——plate 22(all scale bars＝2mm)

1-12. *Cliftonia oxoplecioides* Wright

1. Ventral exterior, DH3, HB669; 2. Ventral internal mold, DH3, HB668; 3. Ventral internal mold, DH2, HB117; 4. Dorsal internal mold, WH3, HB659; 5. Dorsal exterior, DH2, HB72; 6. Dorsal internal mold, WH2, HB675; 7. Dorsal internal mold, WH3, HB658; 8. Ventral exterior, HK1, IV45719; 9. Dorsal internal mold, WH2, HB95; 10. Dorsal internal mold, DH3, HB332; 11. Dorsal internal mold, WH2, HB674; 12. Dorsal internal mold, HH1, HB726。

图版（plate）23

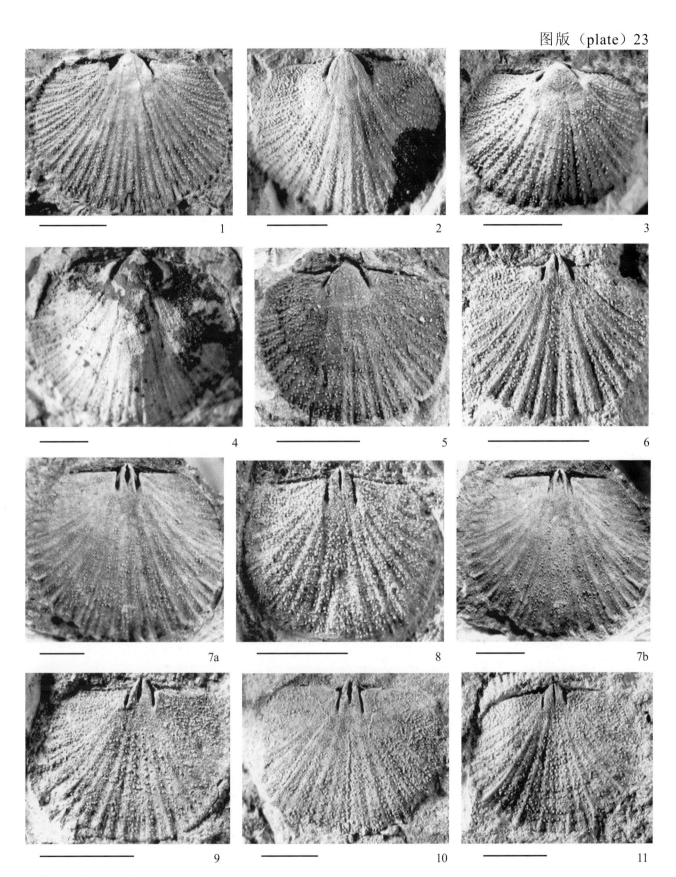

图版 23 所有比例线条＝2mm——plate 23(all scale bars＝2mm)

1-11. *Dalmanella testudinaria* (Dalman)

1. Ventral internal mold, DH2, HB409; 2. Ventral internal mold WH2, HB324; 3. Ventral internal mold, WH2, HB234; 4. Ventral internal mold, DH2, HB184; 5. Ventral internal mold, DH2, HB408; 6. Dorsal internal mold, WH3, HB241; 7a-7b. Same specimen: dorsal internal mold and its cardinalia, DH2, HB339; 8. Dorsal internal mold, WH2, HB246; 9. Dorsal internal mold, WH1, HB418; 10. Dorsal internal mold, WH2, HB244; 11. Dorsal internal mold, WH2, HB245.

图版(plate) 24

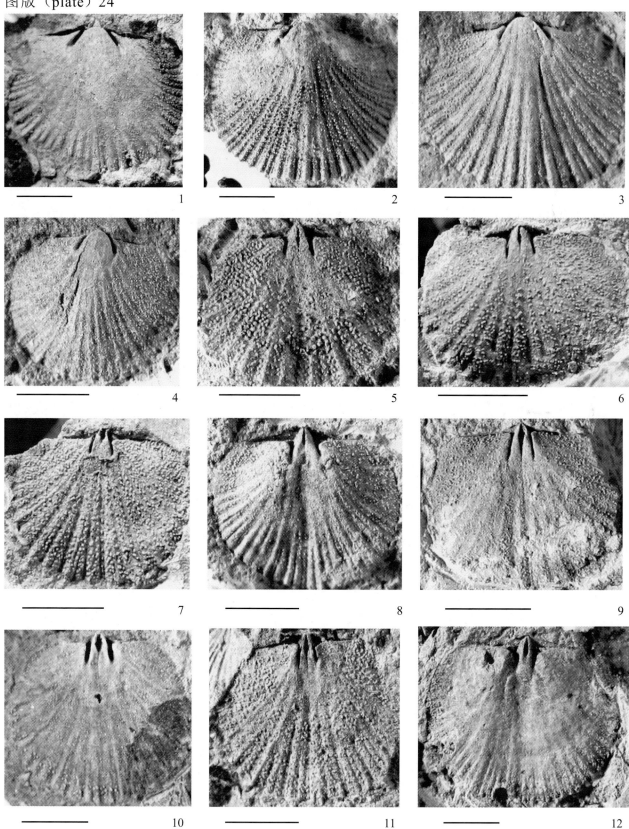

图版 24 所有比例线条=2mm——plate 24(all scale bars=2mm)

1-12. *Dalmanella testudinaria* (Dalman)

1. Ventral internal mold, WH2, HB237; 2. Ventral internal mold, WH1, HB230; 3. Ventral internal mold, WH2, HB243; 4. Ventral internal mold, WH3, HB240; 5. Dorsal internal mold, WH1, HB191; 6. Dorsal internal mold, WH2, HB236; 7. Dorsal internal mold, WH2, HB235; 8. Dorsal internal mold, DH2, HB400; 9. Dorsal internal mold, DH2, HB239; 10. Dorsal internal mold, DH2, HB403; 11. Dorsal internal mold, WH2, HB256; 12. Dorsal internal mold, WH3, HB12.

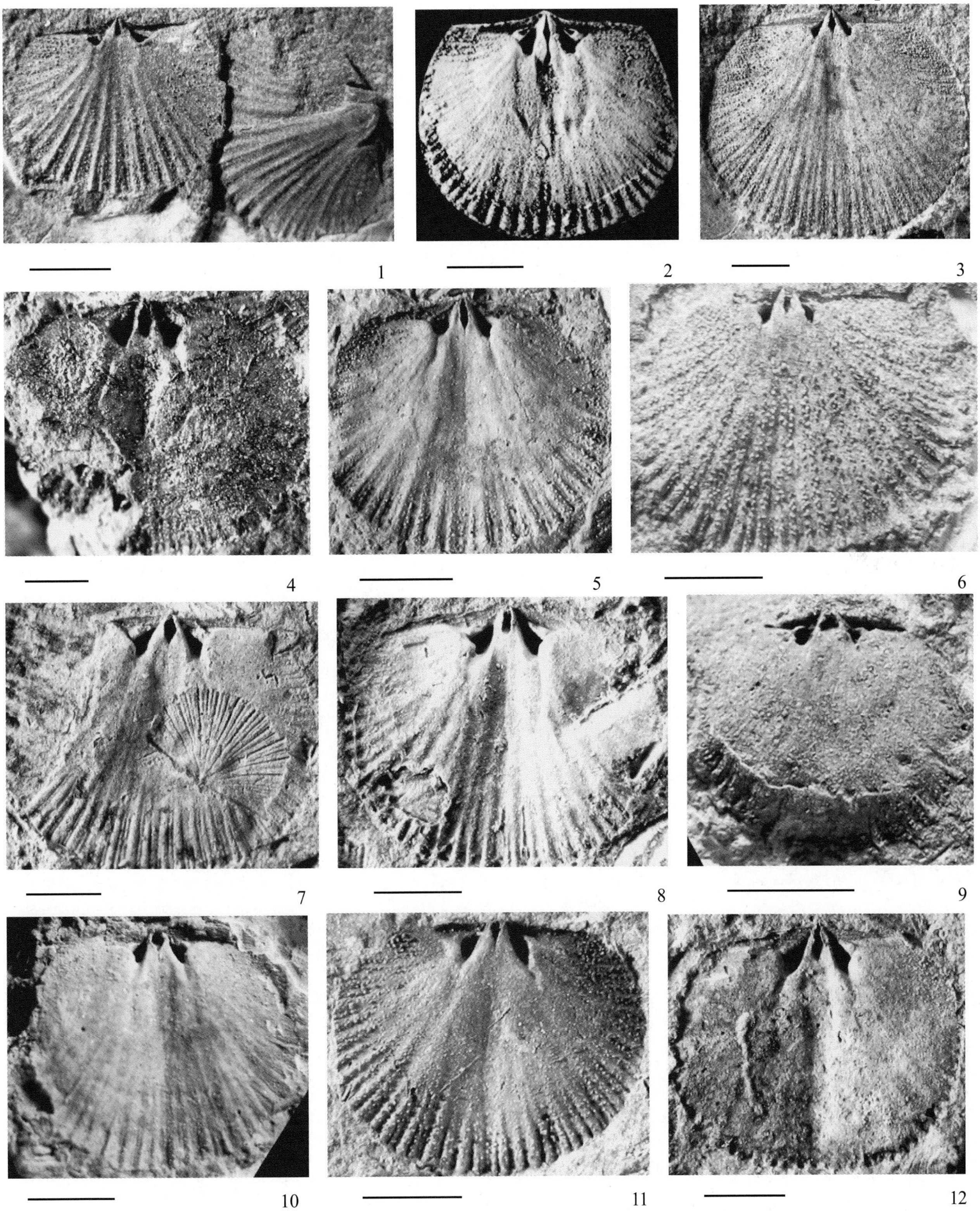

图版（plate）25

图版 25 所有比例线条＝2mm——plate 25(all scale bars＝2mm)

1-12. *Onniella yichangensis* Zeng

1. Dorsal and ventral internal molds, WH1, HB419; 2. Dorsal internal mold, HK1, IV45645; 3. Dorsal internal mold, WH3, HB613-1; 4. Dorsal internal mold, DH2, HB207; 5. Dorsal internal mold, DH2, HB203; 6. Dorsal internal mold, WH3, HB613-2; 7. Dorsal internal mold, DH2, HB202; 8. Dorsal internal mold, DH3, HB513; 9. Dorsal internal mold, HB255; 10. Dorsal internal mold, WH2, HB147; 11. Dorsal internal mold, WH3, HB252; 12. Dorsal internal mold, DH2, HB206.

图版（plate）26

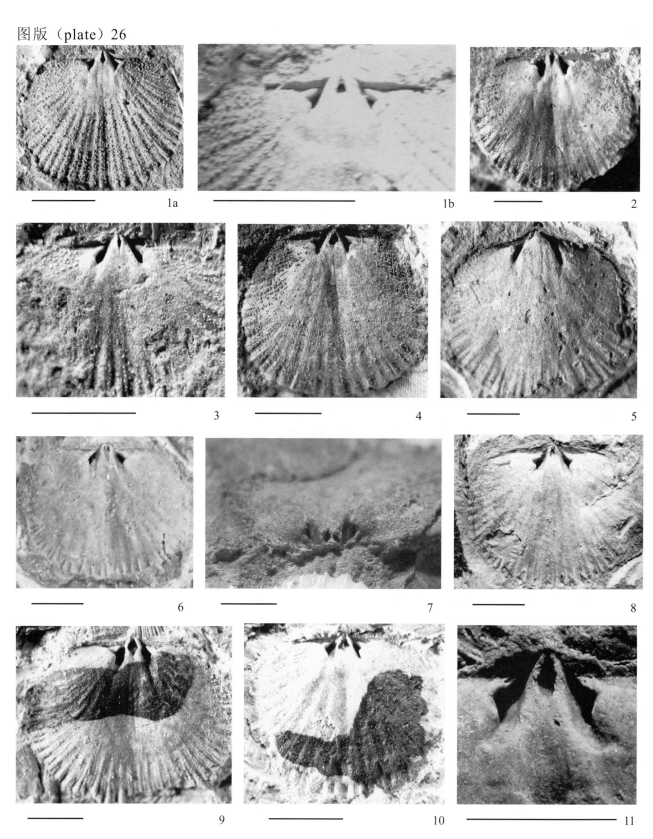

图版 26 所有比例线条＝2mm——plate 26(all scale bars＝2mm)

1-11. *Onniella yichangensis* Zeng

1a-1b. Same specimen: dorsal internal mold and its cardinalia, DH2, HB399; 2. Dorsal internal mold, DH2, HB511; 3. Dorsal internal mold, DH2, HB405; 4. Dorsal internal mold, WH2, HB623; 5. Ventral internal mold, DH2, HB336; 6. Dorsal internal mold, DH3, HB315; 7. Posterior view of transverse section of dorsal beak showing bilobed cardinal process, WH2, HB623a; 8. Dorsal internal mold, DH3, HB513; 9. Dorsal internal mold, DH3, HB334; 10. Dorsal internal mold, DH3, HB524; 11. Cardinalia, DH3, HB102-1.

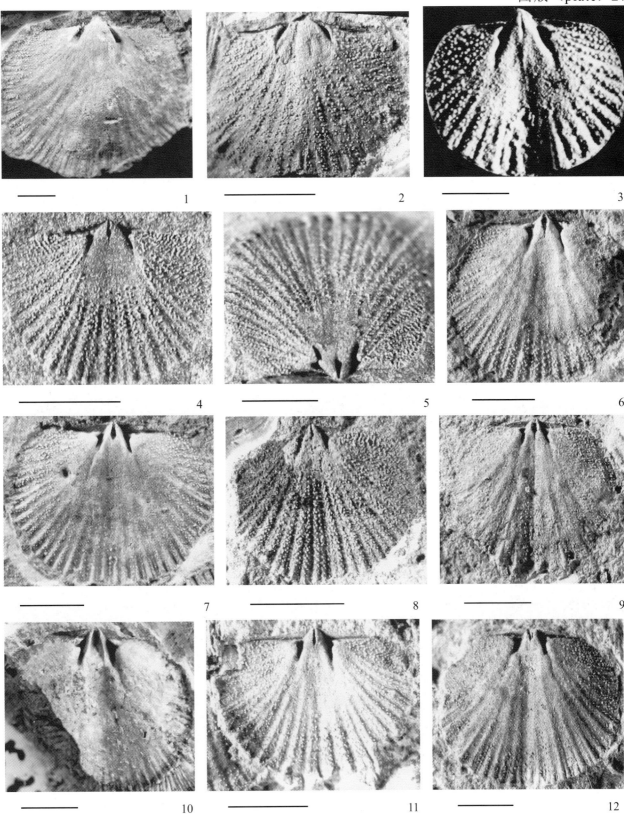

图版 27 所有比例线条=2mm——plate 27(all scale bars=2mm)

1-12. *Trucizetina yichangensis*(Zeng)

1. Ventral internal mold, DH3, HB499; 2. Ventral internal mold, DH2, HB356; 3. Dorsal internal mold, HK3, IV45548; 4. Dorsal internal mold, WH1, HB249; 5. Posterior view of dorsal internal mold, WH1, HB248; 6. Dorsal internal mold, DH3, HB508; 7. Dorsal internal mold, DH3, HB489; 8. Dorsal internal mold, WH2, HB19; 9. Dorsal internal mold, WH2, HB254; 10. Dorsal internal mold, DH3, HB549; 11. Dorsal internal mold, WH2, HB629; 12. Dorsal internal mold, WH2, HB631.

图版（plate）28

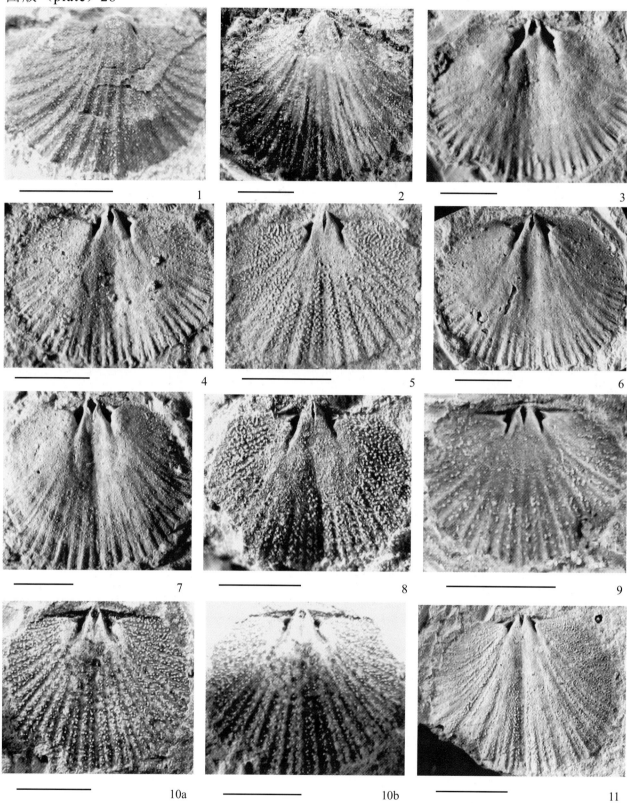

图版 28 所有比例线条＝2mm——plate 28(all scale bars＝2mm)

1-11. *Trucizetina yichangensis*(Zeng)

1. Ventral internal mold, DH3, HB627; 2. Ventral internal mold, DH2, HB407; 3. Dorsal internal mold, DH3, HB115; 4. Dorsal internal mold, DH3, HB141; 5. Dorsal internal mold, DH3, HB626; 6. Dorsal internal mold, DH3, HB151; 7. Dorsal internal mold, WH2, HB147; 8. Dorsal internal mold, WH2, HB617; 9. Dorsal internal mold, DH2, HB550; 10a-10b. Same specimen: Dorsal internal mold and its bilobed cardinal process, DH2, HB555; 11. Dorsal internal mold, WH2, HB618.

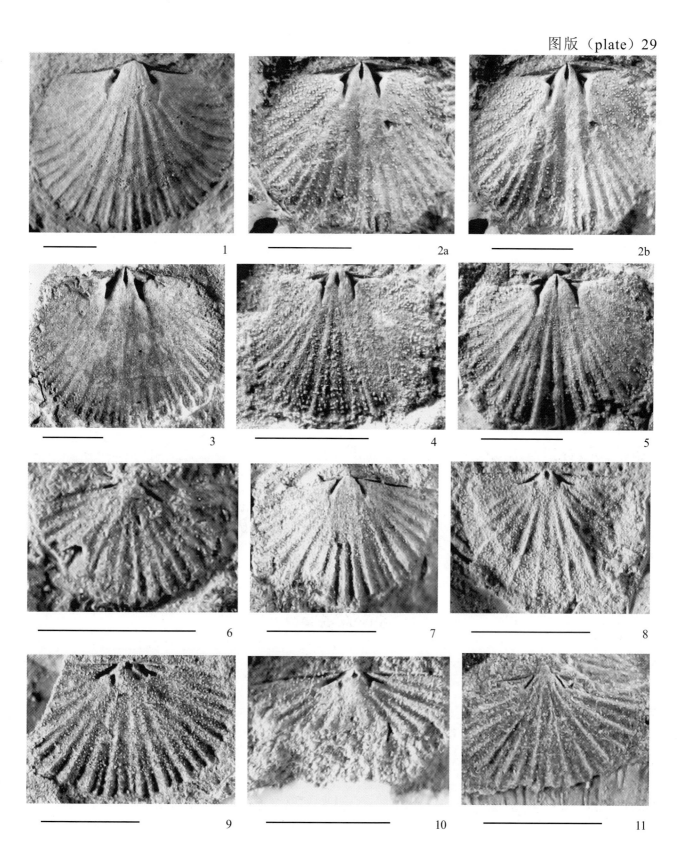

图版 29 所有比例线条=2mm——plate 29(all scale bars=2mm)

1-5. *Trucizetina? parallela* sp. nov.

1. Ventral internal mold, DH3, HB619; 2a-2b. Same specimen: dorsal internal mold and its details of cardinalia, holotype, DH2, HB512;

3. Dorsal internal mold, DH2, HB337; 4. Dorsal internal mold, WH2, HB395; 5. Dorsal internal mold, paratype WH2, HB609.

6-11. *Paramirorthis minuta* gen. et sp. nov.

6. Ventral internal mold, WH2, HB148; 7. Ventral internal mold, WH2, HB187; 8. Dorsal internal mold, DH3, HB140; 9. Dorsal internal mold, WH1, HB170; 10. Dorsal internal mold, WH2, HB169; 11. Ventral internal mold, paratype, DH2, HB174.

图版（plate）30

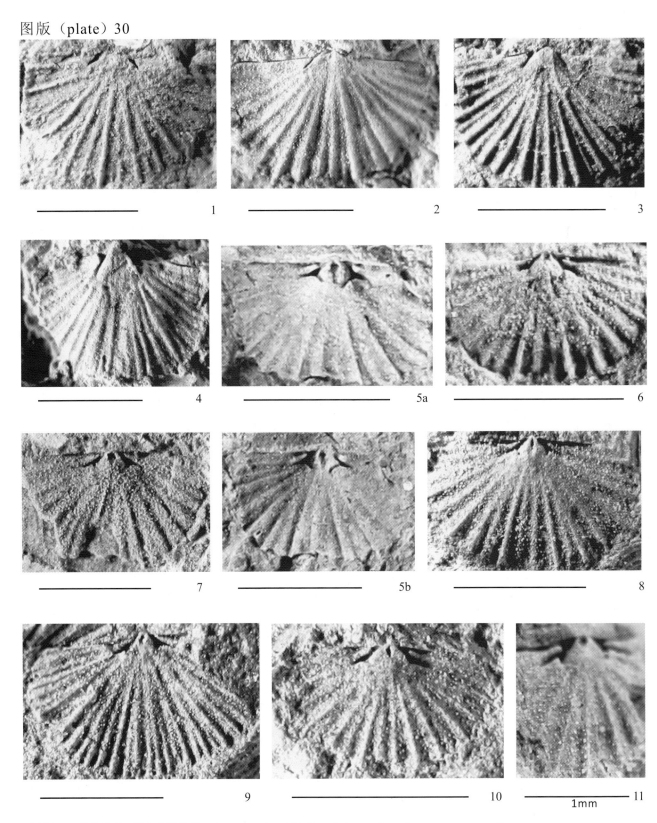

图版 30 除加注外，所有比例线条＝2mm——plate 30(all scale bars＝2mm for except as noted)

1-11. *Paramirorthis minuta* gen. et sp. nov.

1. Ventral internal mold WH2,HB162;2. Ventral internal mold,paratype,WH2,HB157;3. Ventral internal mold,WH2,HB182;4. Ventral internal mold,WH2,HB146;5a-5b. Same specimen:dorsal internal mold,paratype,DH3,HB150;6. Dorsal internal mold,holotype,WH2,HB156;7. Dorsal internal mold,DH2,HB168;8. Dorsal internal mold,paratype,WH2,HB163;9. Dorsal internal mold,WH1,HB171;10. Dorsal internal mold,WH2,HB176;11. Dorsal internal mold,DH2,HB145.

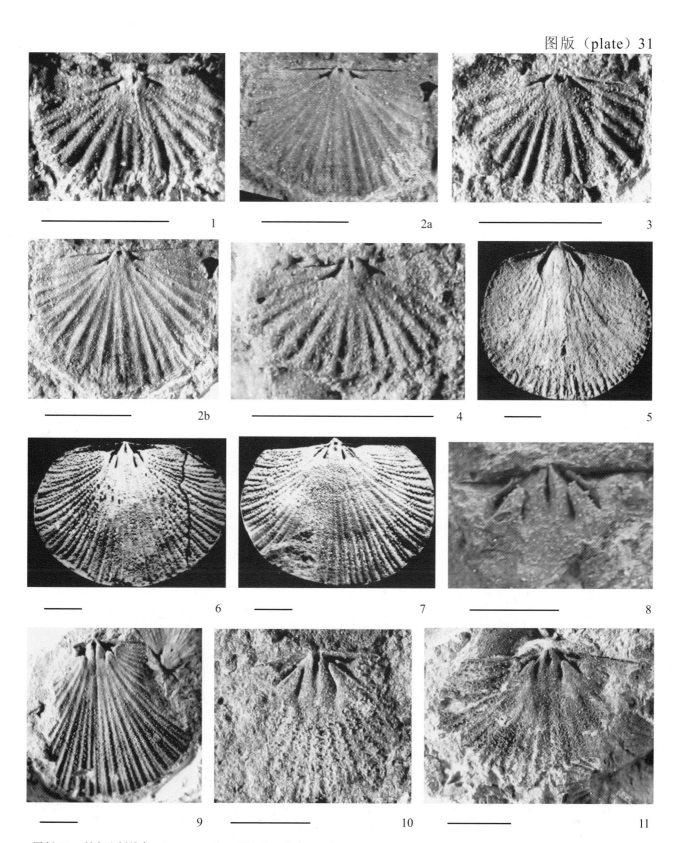

图版 31 所有比例线条＝2mm——plate 31(all scale bars＝2mm)

1－4. *Paramirorthis minuta* gen. et sp. nov.

1. Dorsal internal mold, WH2, HB177; 2a－2b. Same specimen: dorsal internal mold and its ornamentation, paratype, DH2, HB326; 3. Dorsal internal mold, WH1, HB173; 4. Dorsal internal mold, WH2, HB165.

5－11. *Mirorthis mira* Zeng

5. Ventral internal mold, HK3, IV45640; 6. Dorsal internal mold (specimen of original holotype), HK3, IV45639; 7. Dorsal internal mold, HK3, IV45564; 8. Cardinalia, WH1, HB190; 9. Dorsal internal mold, DH3, HB144; 10. Dorsal internal mold, WH2, HB462; 11. Dorsal internal mold, WH2, HB317.

图版（plate）32

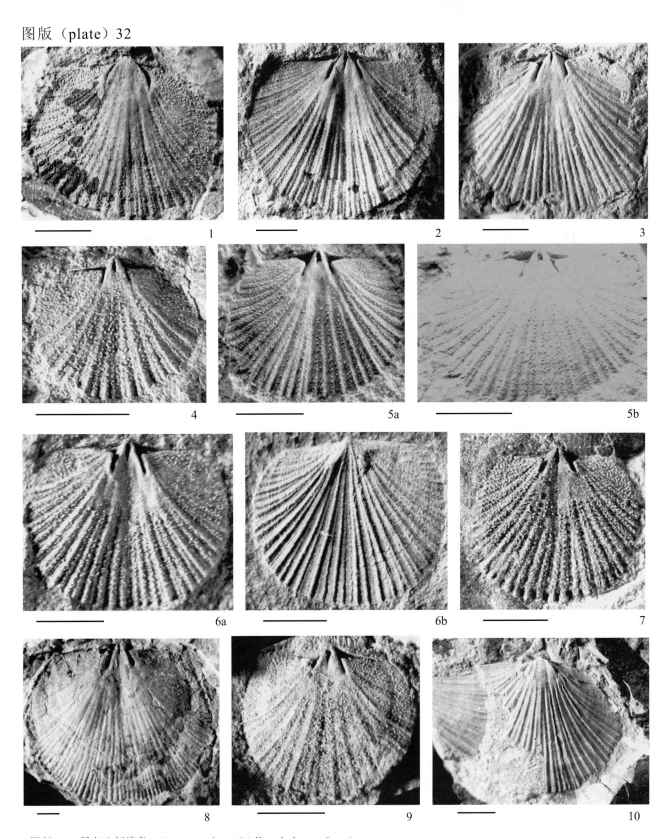

图版 32 所有比例线条＝2mm——plate 32(all scale bars＝2mm)

1-10. *Drabovia dingjiapoensis* sp. nov.

1. Ventral internal mold, WH2, HB676; 2. Dorsal internal mold, DH2, HB542; 3. Dorsal internal mold, DH3, HB153; 4. Dorsal internal mold, WH1, HB231; 5a-5b. Same specimen: dorsal internal mold and its cardinalia, paratype WH3, HB601; 6a-6b. Same specimen: dorsal internal mold and its dorsal external mold, holotype, WH1, HB228; 7. Dorsal internal mold, WH1, HB248; 8. Dorsal internal mold, WH2, HB67; 9. Dorsal internal mold, WH2, HB116; 10. Ventral internal mold, DH2, HB543.

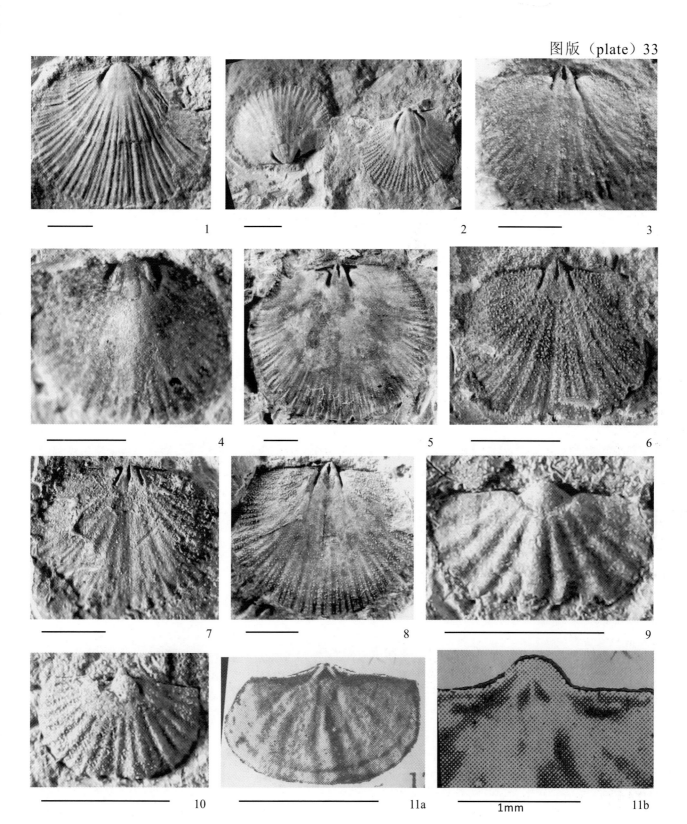

图版33 除加注外,所有比例线条=2mm——plate 33(all scale bars=2mm for except as noted)

1-8. *Draboria dingjiapoensis* sp. nov.

1. Ventral internal mold, DH2, HB554; 2. Ventral internal molds, paratype, DH3, HB559; 3. Dorsal internal mold, DH3, HB533; 4. Ventral internal mold, DH2, HB402; 5. Dorsal internal mold, DH3, HB531; 6. Dorsal internal mold, paratype, WH3, HB695; 7. Dorsal internal mold, DH3, HB299; 8. Dorsal internal mold, DH3, HB397.

9-11. *Toxorthis mirabilis* Rong

9. Ventral internal mold, DH2, HB209; 10. Ventral internal mold, DH2, HB210; 11a-11b. Same specimen: dorsal internal mold and its cardinalia, FK2, IV45653.

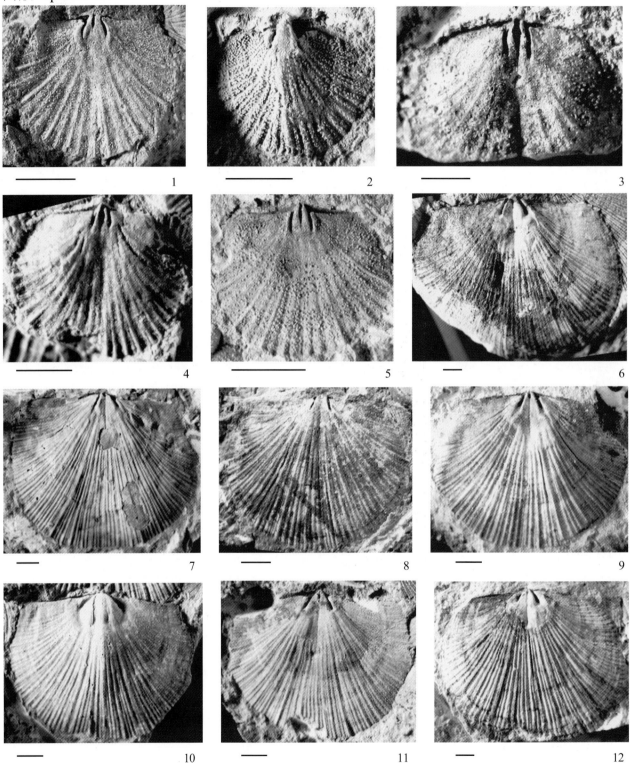

图版 34 所有比例线条＝2mm——plate 34(all scale bars＝2mm)

1. *Drabovia* ? sp.
   1. Dorsal internal mold, DH2, HB189.

2 – 5. *Drabovinella yichangensis* sp. nov.
   2. Ventral internal mold, WH2, HB35; 3. Dorsal internal mold, paratype, DH2, HB74; 4. Dorsal internal mold, WH2, HB53; 5. Dorsal internal mold, holotype, WH2, HB325.

6 – 12. *Hirnantia fecunda* Rong
   6. Dorsal internal mold, WH1, HB46; 7. Dorsal internal mold, DH3, HB712; 8. Dorsal internal mold, DH2, HB553; 9. Dorsal internal mold, DH3, HB514; 10. Ventral internal mold, DH2, HB1; 11. Dorsal internal mold, WH2, HB464; 12. Dorsal internal mold, DH3, HB526.

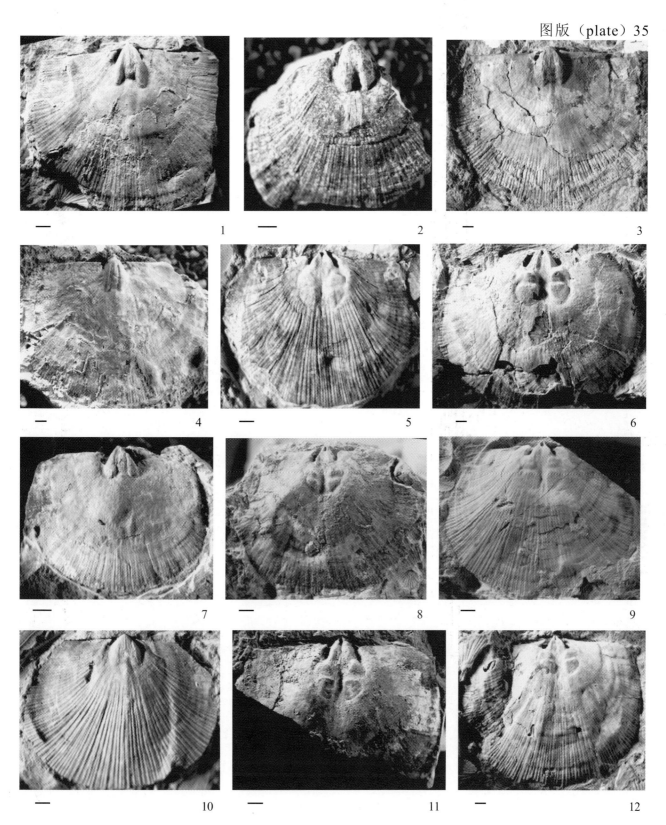

图版 35 所有比例线条＝2mm——plate 35(all scale bars＝2mm)

1-12. *Hirnantia sagittifera* (M'Coy) morph. Bohemia Havliček(1977)

1. Ventral internal mold, WH3, HB126; 2. Ventral internal mold, WH1, HB81; 3. Ventral internal mold, WH2, HB66; 4. Ventral internal mold, WH3, HB51; 5. Dorsal internal mold, WH3, HB9; 6. Dorsal internal mold, WH3, HB131; 7. Ventral internal mold, WH2, HB4; 8. Dorsal internal mold, WH2, HB278; 9. Dorsal internal mold, WH2, HB82; 10. Ventral internal mold, WH3, HB7; 11. Dorsal internal mold, DH2, HB104; 12. Dorsal internal mold, WH3, HB132.

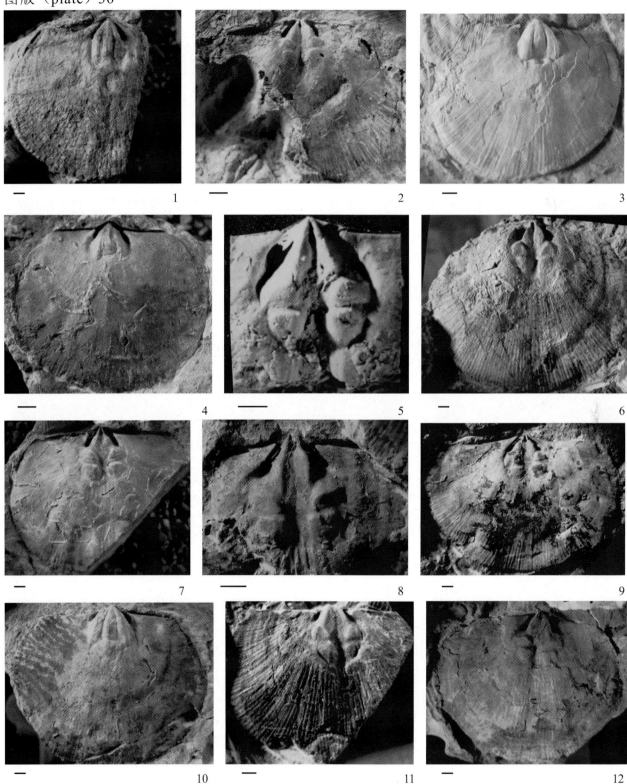

图版36 所有比例线条＝2mm——plate 36(all scale bars＝2mm)

1-2. *Hirnantia sagittifera* (M'Coy) morph. Bohemia Havlíček(1977)

    1. Ventral internal mold, WH2, HB54; 2. Dorsal internal mold, WH3, HB127.

3-12. *Hirnantia magna* Rong

    3. Ventral internal mold, WH3, HB8; 4. Ventral internal mold, WH2, HB7; 5. Dorsal muscle scars, HH2, HB741; 6. Dorsal internal mold, WH2, HB97b; 7. Dorsal internal mold, DH2, HB59; 8. Dorsal internal mold, WH3, HB128; 9. Dorsal internal mold, WH3, HB119; 10. Ventral internal mold, WH3, HB65; 11. Dorsal internal mold, WH1, HB77; 12. Dorsal internal mold, WH3, HB296.

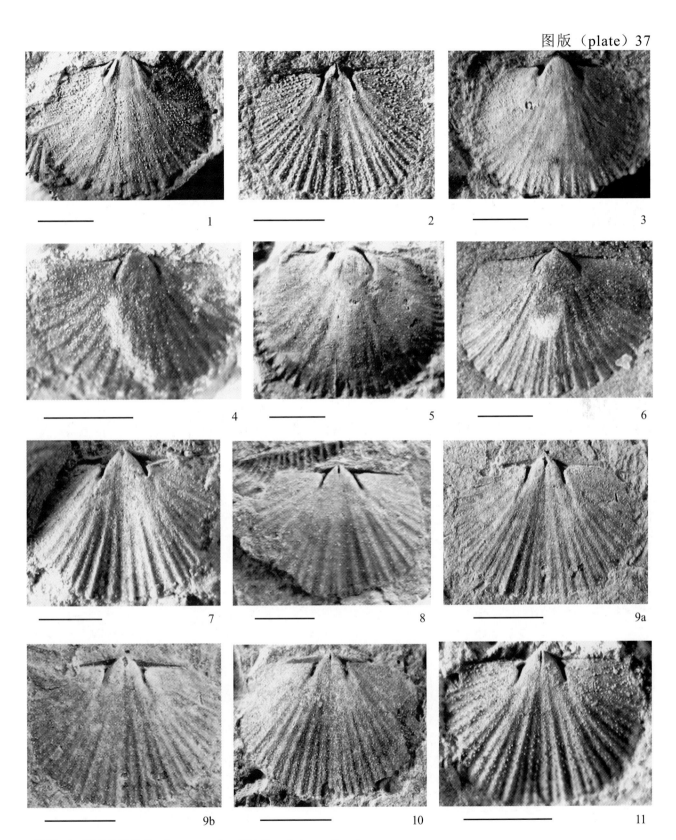

图版 37 所有比例线条＝2mm——plate 37(all scale bars＝2mm)

1-11. *Hirnantia sagittifera* (M'Coy) morph. Poland Temple(1965)

1. Ventral internal mold, DH2, HB204; 2. Dorsal internal mold, WH1, HB200; 3. Ventral internal mold, DH3, HB507; 4. Ventral internal mold, DH3, HB301; 5. Ventral internal mold, DH3, HB396; 6. Ventral internal mold, WH1, HB247; 7. Dorsal internal mold, DH2, HB714; 8. Dorsal internal mold, WH2, HB199; 9a-9b. Sam specimen: dorsal internal mold and its cardinalia, DH3, HB615; 10. Dorsal internal mold, WH3, HB649; 11. Dorsal internal mold, WH2, HB502.

图版（plate）38

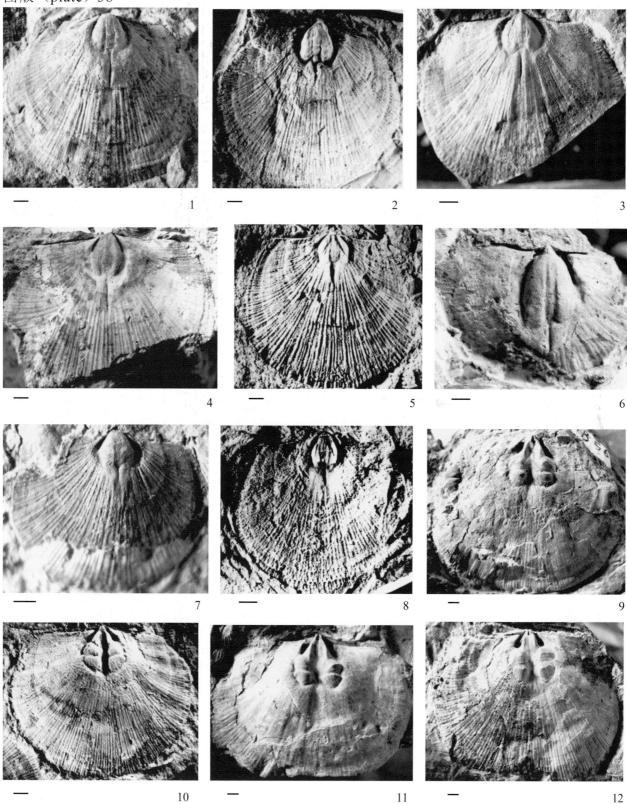

图版 38 所有比例线条＝2mm——plate 38(all scale bars＝2mm)

1-12. *Hirnantia septumis* Zeng

1. Ventral internal mold, WH1, HB85; 2. Ventral internal mold, 花 H2, HB746; 3. Ventral internal mold, DH2, HB84; 4. Ventral internal mold, WH2, HB565; 5. Ventral internal mold, HH2, HB784; 6. Ventral muscle scars, DH2, HB535; 7. Ventral internal mold, WH2, HB80; 8. Ventral internal mold, HH2, HB749; 9. Dorsal internal mold, HB56; 10. Dorsal internal mold, HH2, HB747; 11. Dorsal internal mold, WH1, HB13; 12. Dorsal internal mold, WH3, HB486.

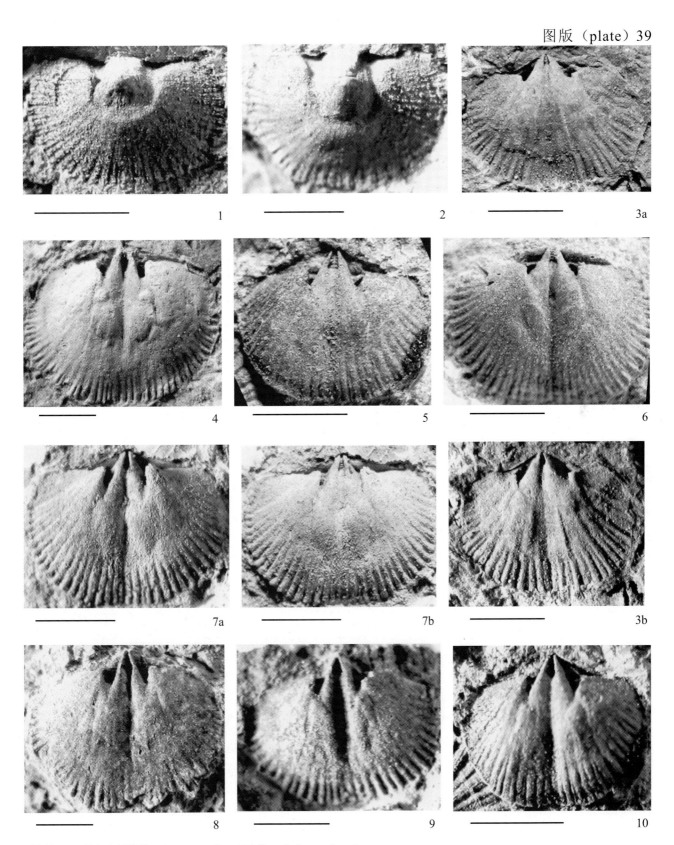

图版 39 所有比例线条＝2mm——plate 39(all scale bars＝2mm)

1-10. *Kinnella kielanae* (Temple)

1. Ventral internal mold, WH2, HB268; 2. Ventral internal mold, WH1, HB262; 3a-3b. Same specimen; showing bilobed cardinal process and dorsal internal mold, DH3, HB30b; 4. Dorsal internal mold, DH3, HB527; 5. Dorsal internal mold, WH1, HB264; 6. Dorsal internal mold, WH3, HB571; 7a-7b. Same specimen; Dorsal internal mold and its details of bilobed cardinal process, DH2, HB401; 8. Dorsal internal mold, WH3, HB273; HB527; 9. Dorsal internal mold, WH2, HB32; 10. Dorsal internal mold, WH1, HB264.

图版（plate）40 所有比例线条＝2mm——plate 40（all scale bars＝2mm）

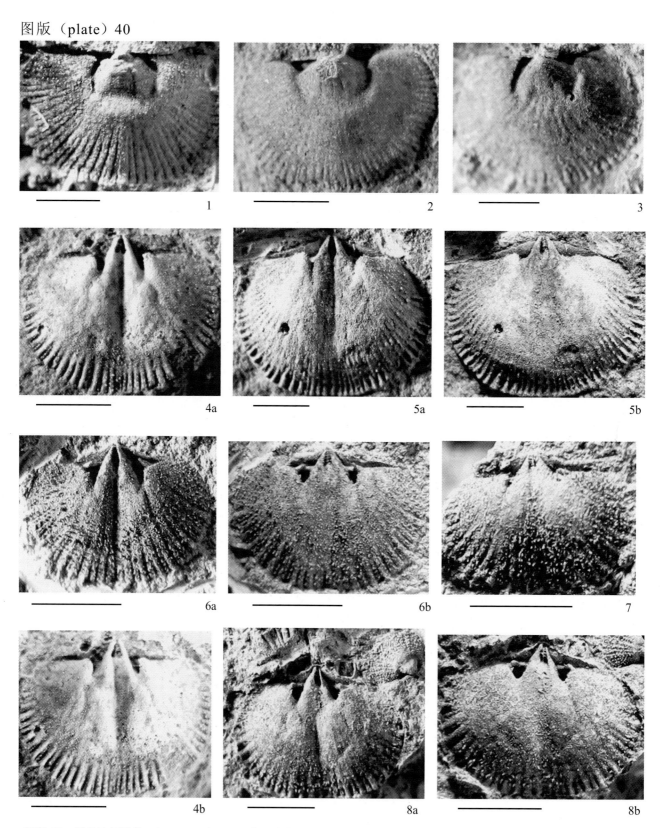

1–8. *Kinnella kielanae* (Temple)

1. Ventral internal mold, WH1, HB258; 2. Ventral internal mold, WH3, HB242; 3. Ventral internal mold, WH2, HB463; 4a–4b. Same specimen: dorsal internal mold and its details of bilobed cardinal process, DH3, HB368; 5a–5b. Same specimen: dorsal internal mold and its details of bilobed cardinal process, WH3, HB259; 6a–6b. Same specimen: dorsal internal mold and its details of bilobed cardinal process, WH1, HB266; 7. Dorsal internal mold, WH1, HB263; 8a–8b. Same specimen: dorsal and ventral internal molds and cardinalia, WH1, HB265.

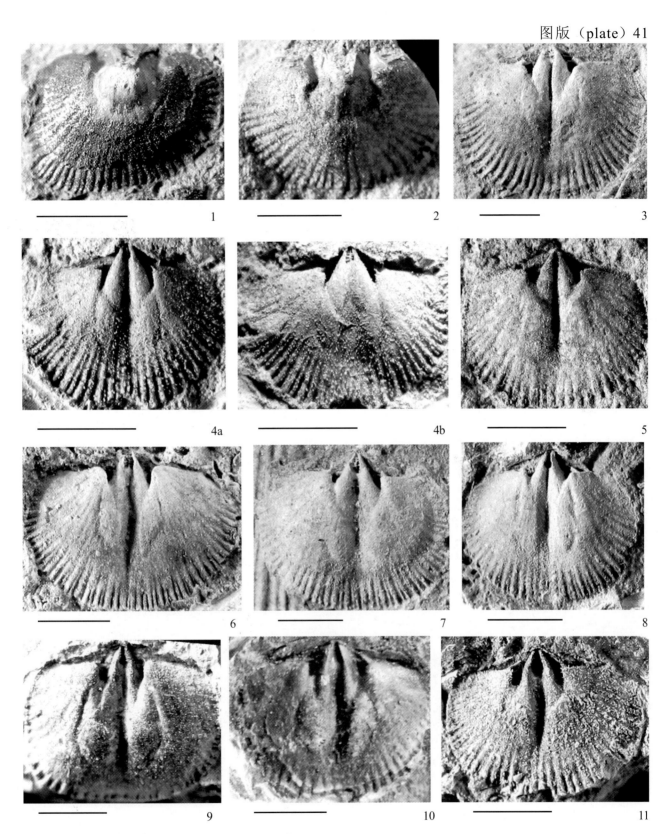

图版41 所有比例线条=2mm——plate 41(all scale bars=2mm)

1-11. *Kinnella robusta* Chang

1. Ventral internal mold, WH2, HB194; 2. Ventral internal mold, WH2, HB269; 3. Dorsal internal mold, DH2, HB494; 4a-4b. Same specimen: dorsal internal mold and its details of bilobed cardinal process, WH2, HB179; 5. Dorsal internal mold, WH2, HB257; 6. Dorsal internal mold, WH2, HB275; 7. Dorsal internal mold, DH2, HB270; 8. Dorsal internal mold, DH2, HB329; 9. Dorsal internal mold, WH3, HB274; 10. Dorsal internal mold, DH3, HB300; 11. Dorsal internal mold, WH2, HB276.

图版（plate）42

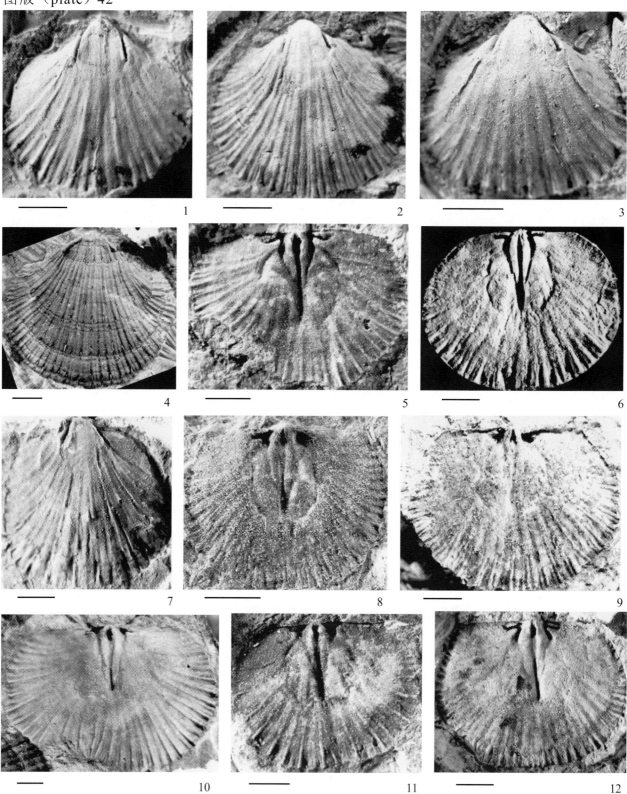

图版 42 所有比例线条＝2mm——plate 42(all scale bars＝2mm)

1–12. *Draborthis caelebs* Marek et Havlíček

1. Ventral internal mold, DH2, HB547; 2. Ventral internal mold, DH2, HB474; 3. Ventral internal mold, DH2, HB475; 4. Ventral external mold, DH3, HB479; 5. Dorsal internal mold, DH3, HB108; 6. Dorsal internal mold, HK3, IV45549; 7. Ventral internal mold, DH3, HB530; 8. Dorsal internal mold, WH3, HB476; 9. Dorsal internal mold, WH3, HB704; 10. Dorsal internal mold, DH3, HB478; 11. Dorsal internal mold, DH3, HB297; 12. Dorsal internal mold, WH2, HB379.

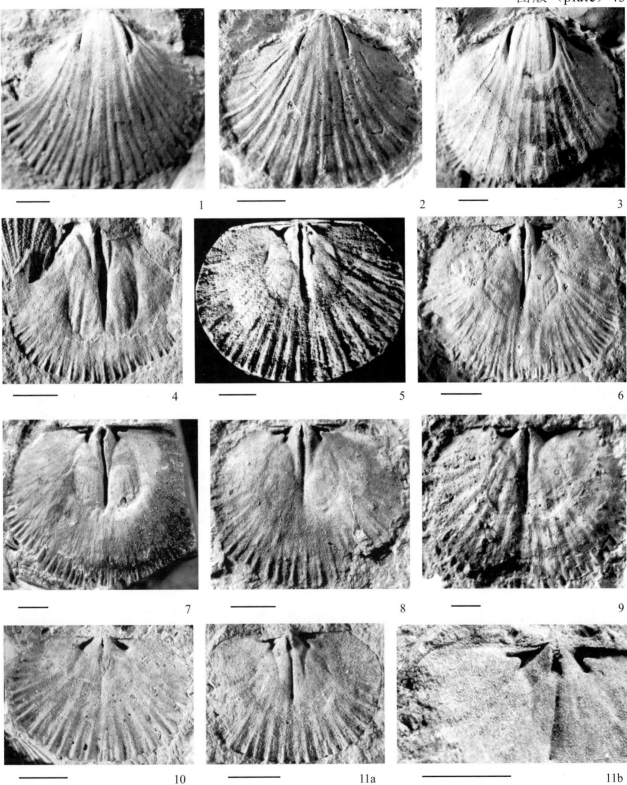

图版 43 所有比例线条=2mm——plate 43(all scale bars=2mm)

1-11. *Draborthis caelebs* Marek et Havlíček

1. Ventral internal mold, DH3, HB109; 2. Ventral internal mold, DH2, HB320; 3. Ventral internal mold, DH2, HB205; 4. Details of dorsal muscle scars, DH3, HB303; 5. Dorsal internal mold, HK3, IV45554; 6. Dorsal internal mold, DH3, HB152; 7. Dorsal internal mold, DH3, HB645; 8. Dorsal internal mold, DH3, HB522; 9. Dorsal internal mold, DH3, HB348; 10. Dorsal internal mold, DH3, HB477; 11a-11b. Same specimen; dorsal internal mold and its local enlargements showing details of cardinalia, WH2, HB472.

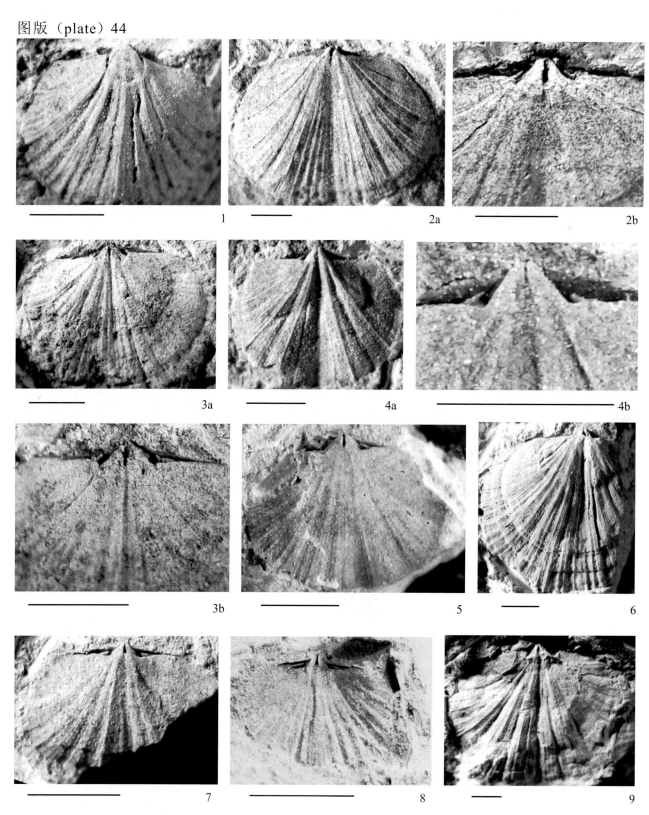

图版 44 所有比例线条＝2mm——plate 44(all scale bars＝2mm)

1-9. *Dysprosorthis sinensis* Rong

1. Ventral internal mold, WH3, HB602; 2a-2b. Same specimen: dorsal internal mold and its local enlargements showing details of cardinalia, WH2, HB211; 3a-3b. Same specimen: dorsal internal mold and its local enlargements showing details of cardinalia, WH2, HB212; 4a-4b. Same specimen: dorsal internal mold and its local enlargements showing details of cardinalia, WH2, HB158; 5. Dorsal internal mold, WH2, HB218; 6. Dorsal internal mold, WH2, HB219; 7. Dorsal internal mold, WH2, HB213; 8. Dorsal internal mold, DH2, HB705; 9. Dorsal internal mold, WH1, HB15.

图版 45 所有比例线条＝2mm——plate 45(all scale bars＝2mm)

1-5. *Cliftonia elongata* sp. nov.
　　1. Ventral internal mold, holotype, HK1, IV45714; 2. Ventral internal mold, HK1, HB724; 3. Dorsal exterior, paratype, HK1, IV45715;
　　4a-4b. Same specimen: dorsal internal mold and its cardinal process, WH1, HB671; 5. Dorsal internal mold, WH2, HB672.

6-11. *Hindella crassa incipiens* (Williams)
　　6. Ventral internal mold, WH2, HB654; 7. Ventral internal mold, HK2, IV45655; 8. Dorsal internal mold, WH1, HB19; 9. Ventral internal mold, WH2, HB661; 10. Dorsal internal mold, HK2, IV45741; 11. Dorsal internal mold, WH3, HB709.

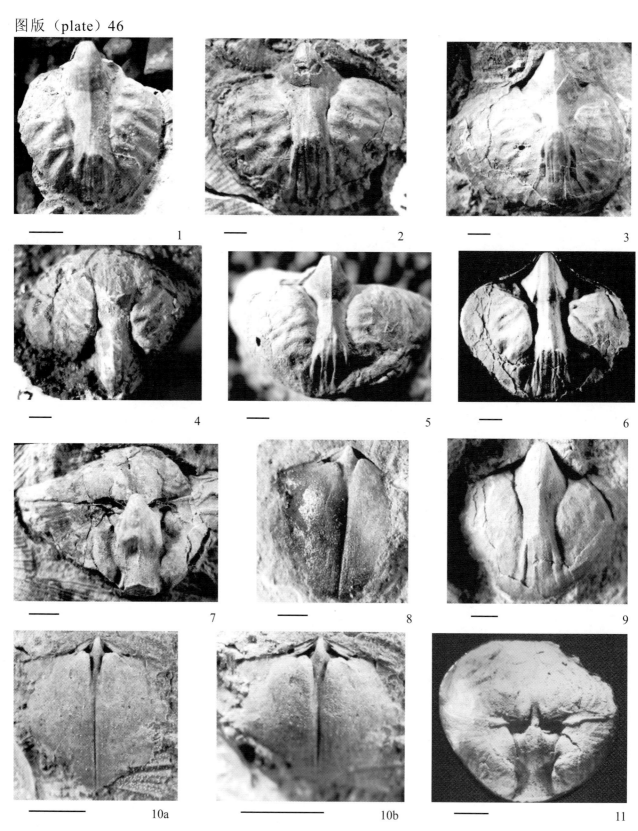

图版 46 所有比例线条＝2mm——plate 46(all scale bars＝2mm)

1-10. *Hindella crassa incipiens* (Williams)

1. Ventral internal mold, WH2, HB663; 2. Ventral internal mold, WH3, HB656; 3. Ventral internal mold, WH3, HB650; 4. Posterior view of ventral internal mold, WH3, HB121; 5. Ventral internal mold, WH2, HB655; 6. Ventral internal mold, HK3, IV45659; 7. Posterior view of internal mold of conjoined valves, WH3, HB603; 8. Dorsal internal mold, DH3, HB333; 9. Ventral internal mold, DH2, HB98; 10a-10b. Same specimen: dorsal internal mold and its details of cardinalia, DH3, HB665.

11. *Hindella? elegans* sp. nov.

11. Posterior view of internal mold of conjoined valves, HK2, IV45657.

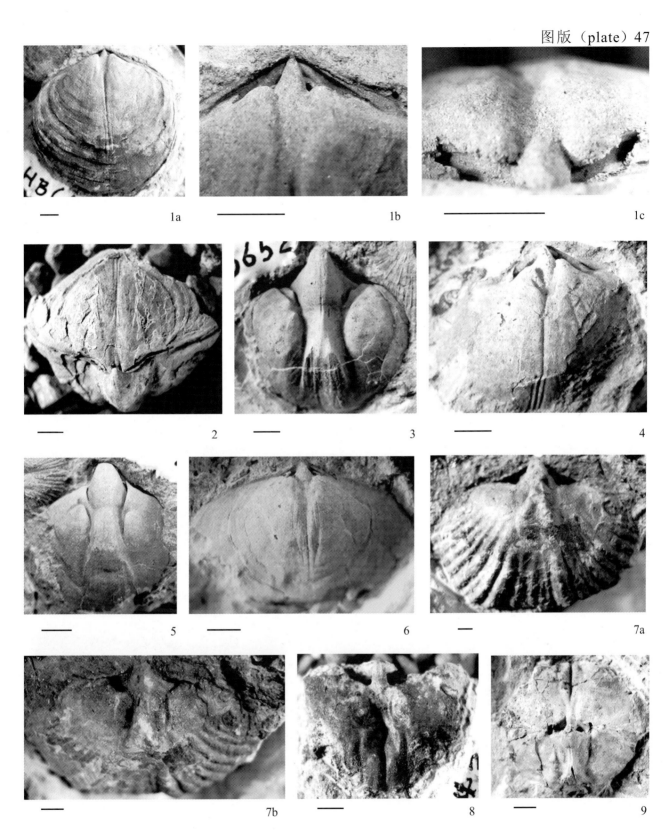

图版 47 所有比例线条＝2mm——plate 47(all scale bars＝2mm)

1-6. *Hindella? elegans* sp. nov.

　　1a-1c. Same specimen: respectively dorsal, cardinalia and posterior views of dorsal internal mold, DH3, HB664; 2. Posterior view of internal mold of conjoined valves, WH2, HB653; 3. Ventral internal mold, holotype, WH2, HB652; 4. Dorsal internal mold, DH3, HB667; 5. Ventral internal mold, DH2, HB352; 6. Dorsal internal mold, paratype, WH2, HB558.

7-9. *Plectothyrella crassicosta* (Dalman)

　　7a-7b. Same specimen: ventral internal mold and its local enlargements showing details of dental plates, DH3, HB639; 8. Dorsal internal mold, WH2, HB643; 9. Posterior view of internal mold of conjoined valves, WH3, HB129.

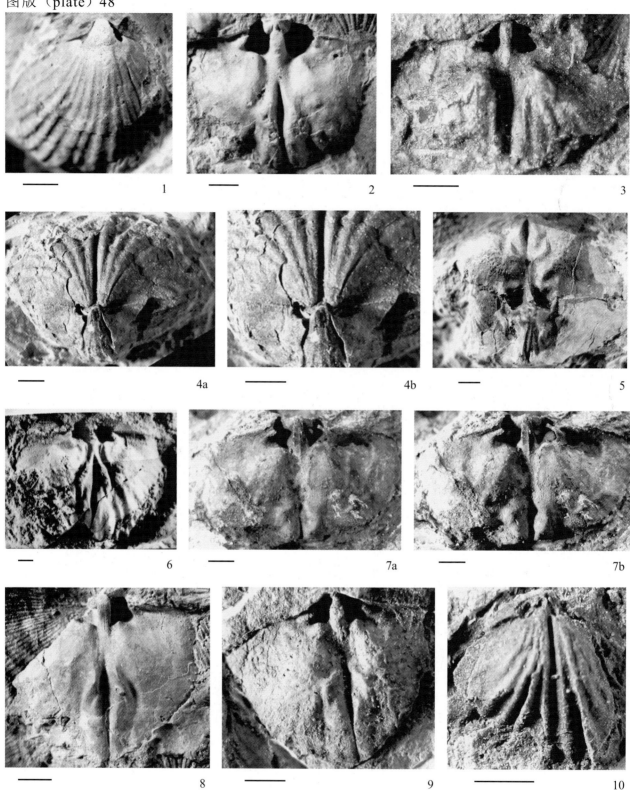

图版 48 所有比例线条=2mm——plate 48(all scale bars=2mm)

1-10. *Plectothyrella crassicosta* (Dalman)

1. Ventral internal mold, DH3, HB711; 2. Dorsal internal mold, DH3, HB107; 3. Dorsal internal mold, WH2, HB642; 4a-4b. Same specimen: Posterior views of internal mold of conjoined valves and its local enlargements showing details of interior structures of ventral and dorsal valves, WH3, HB136; 5. Posterior view of internal mold of conjoined valves, WH2, HB641; 6. Dorsal internal mold, FH2, H748; 7a-7b. Same specimen: dorsal internal mold and its cardinalia, WH3, HB644; 8. Dorsal internal mold, WH2, HB684; 9. Dorsal internal mold, DH3, HB298; 10. Dorsal exterior, WH2, HB683.

图版（plate）49

图版 49 所有比例线条＝2mm——plate 49(all scale bars＝2mm)

1. *Leptaena huanghuaensis* Zeng

    Bilobed cardinal process and secondary sockets of dorsal internal mold, DH3, HB305b.

2. *Leptaenopoma trifidum* Marek et Havlíček

    Posterior view of dorsal internal mold showing strongly dorsal platform and trifided cardinal process, DH3, HB310.

图版（plate）50

1

2

图版 50　所有比例线条＝2mm——plate 50(all scale bars＝2mm)

1-2. *Eostropheodonta hirnantensis* (M'Coy)

　　1. Dental plate denticulates, DH2, HB536; 2. Bilobed cardinal process and denticles on inner socket ridge, WH2, HB578.

图版 (plate) 51

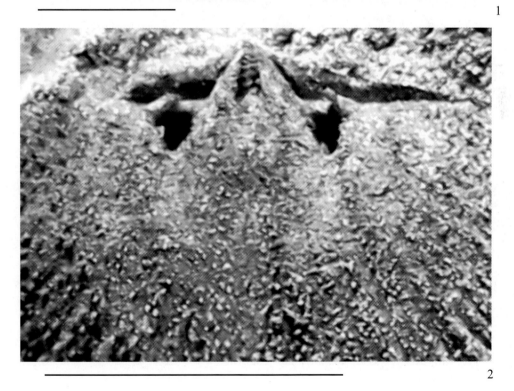

图版 51 所有比例线条＝2mm——plate 51(all scale bars＝2mm)

1. *Aphanomena parvicostellata* Rong
   Details of cardinalia, WH2, HB44.
2. *Kinnella kielanae* (Temple)
   Crenulations on bilobed cardinal process, WH1, HB266.

图版（plate）52

图版52 所有比例线条＝2mm——plate 52(all scale bars＝2mm)

1-2. *Yichangomena dingjiapoensis* gen. et sp. nov.

1. Dental plate denticulates, DH2, HB544. 2. Dental plate denticulates, DH2, HB520.

图版（plate）53

1

2

图版53 所有比例线条＝2mm——plate 53(all scale bars＝2mm)

1-2. *Sinomena typical* gen. et sp. nov.

    1. Dental plate denticulates, WH2, HB607. 2. Dorsal internal mold, WH2, HB563.

图版（plate）54

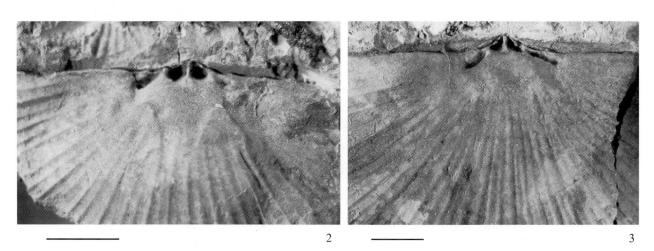

图版 54 所有比例线条＝2mm——plate 54(all scale bars＝2mm)

1-3. *Hubeinomena wangjiawanensis* gen. et sp. Nov.

    1. Dorsal internal mold, showing denticles on inner socket ridge, WH3, HB556.

    2. Details of cardinalia and dorsal muscle scars, WH3, HB696.

    3. Details of cardinalia, WH3, HB660.

图版（plate）55

图版55 所有比例线条＝2mm——plate 55(all scale bars＝2mm)

1-2. *Dysprosorthis sinensis* Rong

1. Dorsal internal mold,WH2,HB212.
2. Dorsal internal mold and its local enlargements showing details of cardinalia,WH2,HB158.

图版（plate）56

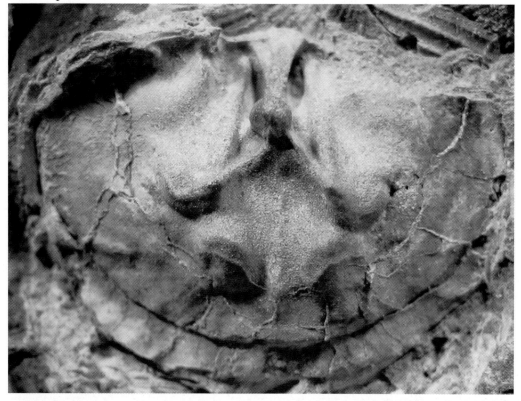

图版 56 所有比例线条＝2mm——plate 56(all scale bars＝2mm)

1-2. *Cliftonia oxoplecioides* Wright
    1. Dorsal muscle scars, WH3, HB659.
    2. Dorsal internal mold, WH2, HB674.

图版（plate）57

1

2

图版 57　所有比例线条＝2mm——plate 57(all scale bars＝2mm)

1. *Paramirorthis minuta* gen. et sp. nov.

　　1. Dorsal internal mold, WH2, HB156.

2. *Mirorthis mira* Zeng

　　2. Dorsal internal mold(specimen of original holotype), HK3, IV45639.

图版（plate）58

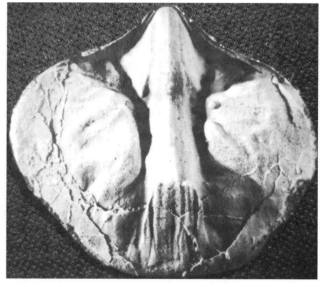

图版 58　所有比例线条＝2mm——plate 58(all scale bars＝2mm)

1-3. *Hindella crassa incipiens* (Williams)

    1. Posterior view of ventral internal mold, WH3, HB121.

    2. Ventral internal mold, HK3, IV45659.

    3. Details of cardinalia, DH3, HB665.

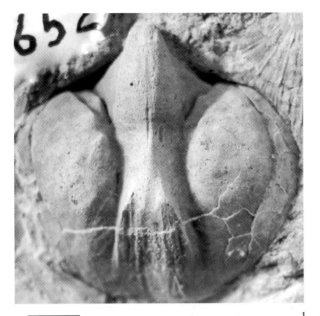

图版 59 所有比例线条＝2mm——plate 59(all scale bars＝2mm)

1－3. *Hindella? elegans* sp. nov.

    1. Ventral internal mold, WH2, HB652.

    2. Posterior view of internal mold of conjoined valves, HK2, IV45657.

    3. Dorsal internal mold, WH2, HB558.

图版（plate）60

图版 60 所有比例线条＝2mm——plate 60(all scale bars＝2mm)

1-2. *Plectothyrella crassicosta* (Dalman)
　　1. Dorsal internal mold, DH3, HB107.
　　2. Details of cardinalia, WH3, HB644.

图版（plate）61

图版61 所有比例线条＝2mm——plate 61(all scale bars＝2mm)

1.宜昌丁家坡五峰组观音桥段中部,即化石采集号DH2地层段浊流沉积层序剖面:A.恢复为正常环境的化石埋藏层;B.由DH2层埋藏的介壳化石遭浊流搬运再沉积的介壳碎片层;C.粒度由粗变细的粒度递变层(gradedbed);D.浊流(湍流—turbulent)运动刻蚀浊流之前的沉积层,并挖成槽模(flute casts);E.产生浊流之前的沉积层(五峰组笔石页岩段最顶部硅质页岩)。

2.宜昌王家湾五峰组观音桥段中部浊流沉积层序剖面,其浊流沉积层序同本图版图1的浊流沉积层序相同,但与图1剖面不同的是,浊流刻蚀槽模的层位要比图1剖面高12～15cm,即位于WH2地层段的中部。

图版（plate）62

1

2

图版62 所有比例线条=2mm——plate 62(all scale bars=2mm)

    1. 为本书图版61,图2的C层颗粒递变层中的石英颗粒夹层(厚度很不均一,通常为1.5cm),石英粒径0.1～0.99mm,但以0.1～0.25mm为多,分选不好,呈棱角状至半圆形,部分为碎玻璃状,个别有火山玻璃质包体,为火山晶屑。

    2. 为本书图版61,图2的B层,即经浊流搬运再沉积的介壳碎片层。

图版（plate）63

图版 63　比例线条＝2mm——plate 63(scale bar＝2mm)

宜昌王家湾五峰组观音桥段顶与龙马溪组底之间的转换分界面(Change-boundary Plane)(即龙马溪组底面)Hirnantia fauna 的成员 Dalmanella sp.，Drabovia sp.，以及其他成员的介壳与龙马溪组最下部笔石带 Normalograptus persculptus 带的成员共同保存在很平整的转换分界面上的情况。

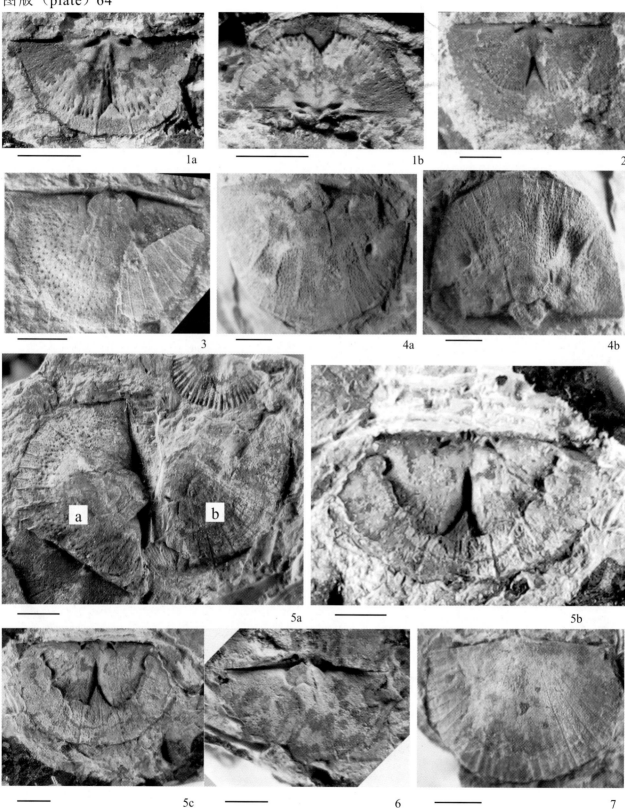

图版 64 所有比例线条＝2mm——plate 64(all scale bars＝2mm)

1-3. *Leptellina(Merciella) striata* Rong, Xu et Yang.
 1a-1b. Same specimen; 1a. dorsal internal mold, pm065-6-1F, YB3; 1b. posterior view. 2. Dorsal internal mold, pm065-4-1F, YB6; 3. Ventral internal mold, pm065-6-1F, YB10.

4-7. *Leptelloidea silurica* sp. nov.
 4a-4b. Same specimen; 4a. ventral internal mold, paratype, pm065-6-1F, YB4; 4b. posterior view; 5a-a. Ventral internal mold, pm065-6-1F, YB7; 5a-b and 5b-5c. Same specimen; 5a-b. dorsal external mold; 5b-5c. dorsal internal molds of 5a-b, holotype; 6. Ventral internal mold, pm065-6-1F, YB9; 7. Ventral exterior, pm065-6-1F, YB73.

图版（plate）65

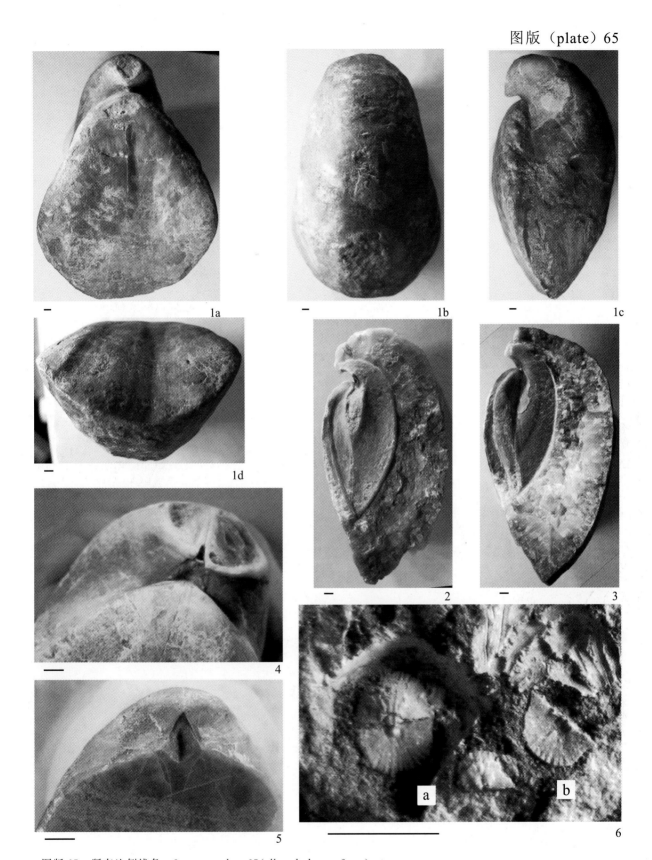

图版65 所有比例线条＝2mm——plate 65(all scale bars＝2mm)

1-5. *Sulcipentamerus dorsoplanus* (Wang)

1a-1d. Respectively dorsal, ventral, lateral, anterior views, pm064-18层, YB74; 2. Lateral view of interior (longitudinal section), pm064-18层, YB77; 3. Lateral view of interior, pm064-18层, YB78; 4. Transverse section, showing spondylium and a concave deltidium of rhomboidal outline, pm064-18层, YB81; 5. Transverse section, showing spondylium and a concave deltidium of rhomboidal outline, pm064-18层, YB76.

6. *Minutorthis yichangensis* (Rong) gen. nov.

6-a. Internal mold of conjoined valves, paratype, pm065-7-2F, YB82; 6-b. Dorsal interior, holotype.

图版（plate）66

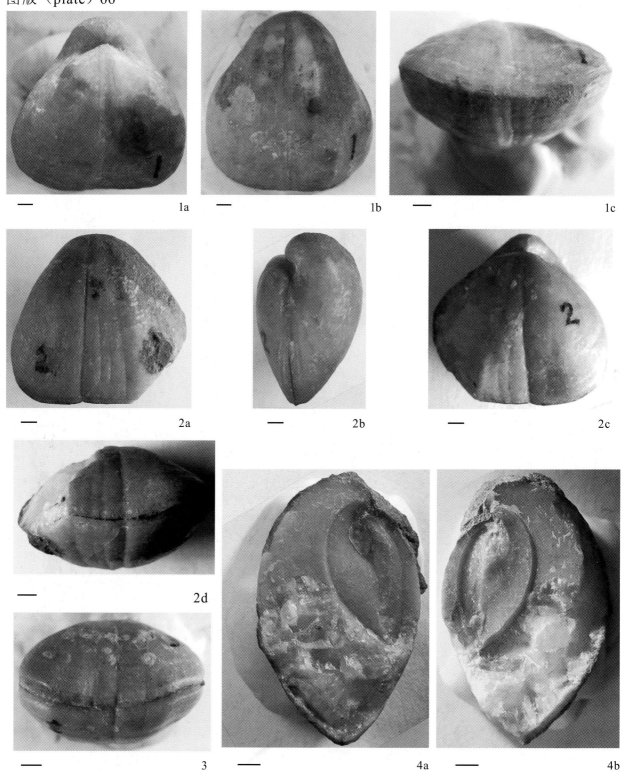

图版 66 所有比例线条＝2mm——plate 66(all scale bars＝2mm)

1-4. *Centreplicatus triangulatus* gen. et sp. nov.

　　1a-1c. Respectively dorsal, ventral, anterior views, paratype, Jl-1, HB720; 2a-2d. Respectively ventral, lateral, dorsal and anterior views, holotype, Jl-1, HB721; 3. Anterior view, Jl-1, HB722; 4a-4b. Lateral views of interior(longitudinal section), paratype, Jl-1, HB727.

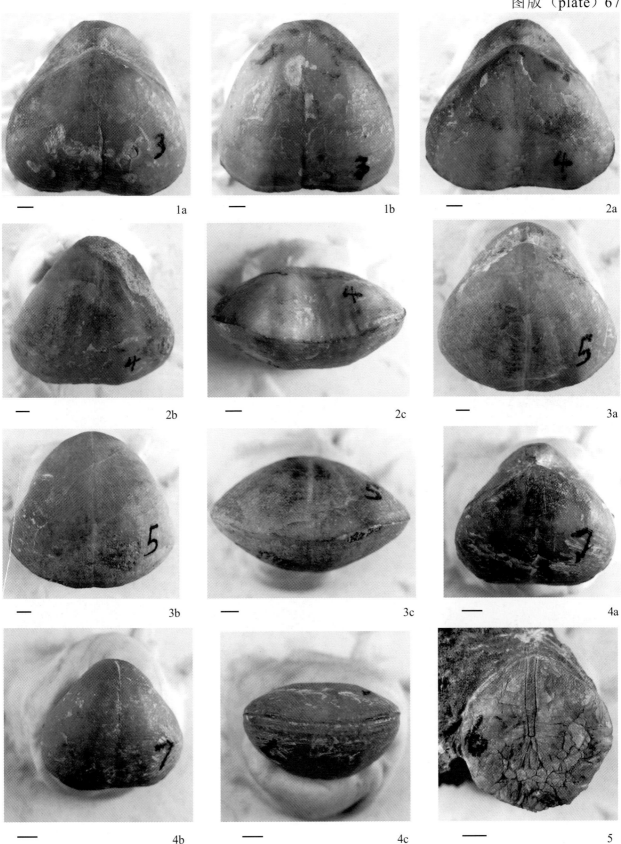

图版67 所有比例线条＝2mm——plate 67 (all scale bars＝2mm)

1–5. *Centreplicatus triangulatus* gen. et sp. nov.

1a–1b. Dorsal and ventral views, Jl–1, HB722; 2a–2c. Respectively dorsal, ventral, anterior views, Jl–1, HB273; 3a–3c. Respectively dorsal, ventral and anterior views, Jl–1, HB274; 4a–4c. A junior valve, respectively dorsal, ventral and anterior views, Jl–1, HB726; 5. The shape of inner hinge plates on weathering front of conjoined valves, paratype, Jl–1, HB725.

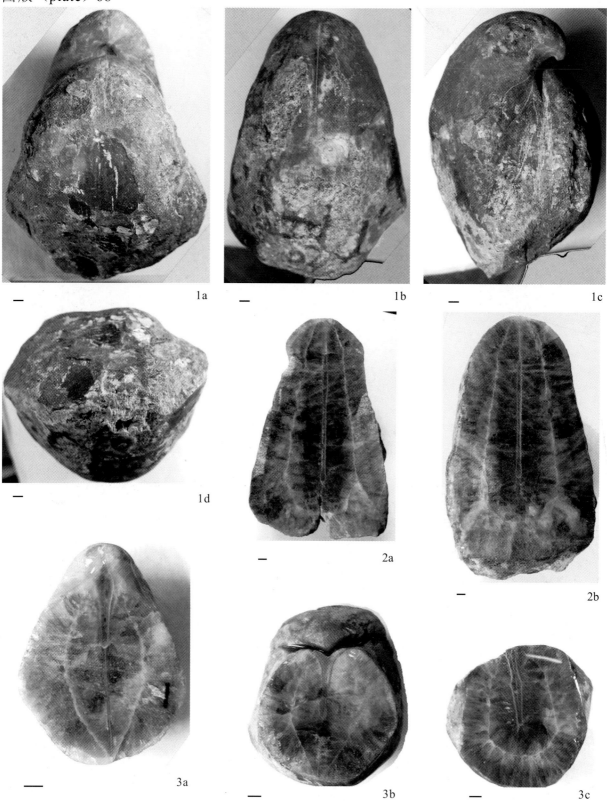

图版 68 所有比例线条＝2mm——plate 68(all scale bars＝2mm)

1-3. *Apopentamerus*(*Enclosurus*) *fenxiangensis*(Zeng)subgen. nov.

1a-1d. Respectively dorsal, ventral, lateral and anterior views, holotype, pm064-18层, YB80; 2a-2b. Same speciment paratype: lateral sections, showing spondylium and calcific enclosures in the ventral chambers, pm064-18层, YB75; 3a-3c. Serial sections of a mature shell, paratype, showing deltidium and spondylium, median septum, calcific enclosures, outer hinge plate, brachial process, inner hinge plate in the visceral cavity, pm064-18层, YB79.